Positive Plant Interactions and Community Dynamics

Positive Plant Interactions and Community Dynamics

Edited by Francisco I. Pugnaire

CRC Press
Taylor & Francis Group
Boca Raton London New York

CRC Press is an imprint of the
Taylor & Francis Group, an **informa** business

Fundación **BBVA**

Fundación BBVA
Plaza de San Nicolás, 4.
48005 Bilbao, Spain
www.fbbva.es

CRC Press
Taylor & Francis Group
6000 Broken Sound Parkway NW, Suite 300
Boca Raton, FL 33487-2742

First issued in paperback 2017

© 2010 by Fundación BBVA
CRC Press is an imprint of Taylor & Francis Group, an Informa business

No claim to original U.S. Government works

ISBN-13: 978-1-4398-2494-8 (hbk)
ISBN-13: 978-1-138-11633-7 (pbk)

Library of Congress Cataloging-in-Publication Data

Positive plant interactions and community dynamics / editor, Francisco I. Pugnaire. --
    1st ed.
        p. cm.
    "The contents of this volume were drafted at a workshop on positive plant interactions held in Madrid and sponsored by the BBVA Foundation"--Pref.
    Includes bibliographical references and index.
    ISBN 978-1-4398-2494-8 (alk. paper)
    1. Plant communities. 2. Plant ecology. I. Pugnaire, Francisco I., 1957-

QK911.P67 2010
581.7'82--dc22   2009032042

Visit the Taylor & Francis Web site at
http://www.taylorandfrancis.com

and the CRC Press Web site at
http://www.crcpress.com

Contents

Preface

The contents of this volume were drafted at a workshop on positive plant interactions held in Madrid and sponsored by the BBVA Foundation. It gathers a compilation of papers in which the authors analyze different aspects of plant–plant interactions, emphasizing the role of facilitation. The study of associations, interactions, and gradients has led to a tremendous growth in community ecology, and the importance of both competition and facilitation has been increasingly acknowledged as two sides of the same coin, even though competition has traditionally received more attention.

The primary goal of this volume is to introduce readers to the application of interaction theory. The topics treated here comprise the bulk of research on plant interactions carried out in the last 10 years. For this book, we chose to develop plant–plant interactions thoroughly while omitting plant–pollinator and plant–herbivore interactions, which are best treated separately and are themselves major avenues of research. Likewise, aspects such as the use of facilitation in ecosystem management and restoration, or the role of indirect interactions, would also need further attention.

I hope this highly interesting set of papers will appeal to scientists working in plant community dynamics, managers, students and, to some extent, the general public. Complex processes like those addressed here are at the base of ecosystem functioning and need to be well understood to properly manage natural habitats, prevent environmental risks such as invasions, and secure the continued delivery of ecosystem services to society.

I would like to thank once again the BBVA Foundation for its efforts in support of biodiversity conservation and the development of biodiversity science. I am indebted to the contributors of this volume for their eagerness, patience, and friendship. Thanks also to the people who provided comments and reviews of the different manuscripts, helping improve their contents. Finally, thanks to CRC Press, Taylor & Francis Group, and in particular, to John Sulzycki for his enthusiastic support of the publication of this book.

Francisco I. Pugnaire

Editor

Francisco I. Pugnaire, Ph.D., is a research scientist who has been with the Spanish National Council for Scientific Research at the Arid Zones Experimental Station (Almería) since 1996. After receiving a Ph.D. from the University of Granada (1989), he did postdoctoral work at the University of California-Berkeley, the University of Complutense (Madrid), and Leeds University (UK). His research interests focus on physiological and functional plant ecology, particularly water relations and gas exchange, species interactions and community dynamics, as well as their applications to ecosystem restoration. Although his work is centered mainly on semi-arid environments, he has worked also in desert ecosystems, tropical forests, and high mountains.

An author of research articles, reviews, and published papers, Dr. Pugnaire has participated in national and international research projects. He is a reviewer for many scientific journals and a member of the editorial teams of *Ecography*, *The New Phytologist*, and *Ecosistemas*. He is responsible for *Web Ecology*, the journal of the European Ecological Federation. A regular member of panels in the Spanish Ministry of Education and Science, he was manager of the Spanish National Programme for International Cooperation and co-manager of the Spanish Programme on Biodiversity and Global Change. He is president of the Diversitas Spanish Committee. He is vice president of the European Ecological Federation and has served on the board of the Spanish Ecological Society.

Contributors

Cristina Armas
Estación Experimental de Zonas
  Áridas
Consejo Superior de Investigaciones
  Científicas
Almería, Spain

Ernesto I. Badano
Departamento de Ecología
Facultad de Ciencias Biológicas
Pontificia Universidad Católica
Santiago, Chile
and
Instituto de Ecología y Biodiversidad
Santiago, Chile

Rob W. Brooker
The Macaulay Institute
Craigiebuckler, Aberdeen
United Kingdom

Ragan M. Callaway
Division of Biological Sciences
The University of Montana
Missoula, Montana, U.S.A.

Lohengrin A. Cavieres
Departamento de Botánica
Facultad de Ciencias Naturales y
  Oceanográficas
Universidad de Concepción
Concepción, Chile
and
Instituto de Ecología y Biodiversidad
Santiago, Chile

Zaal Kikvidze
School of Environmental Sciences
The University of Tokyo
Kashiwanoha, Kashiwa
Chiba, Japan

Christopher J. Lortie
Department of Biology
York University
Toronto, Ontario, Canada

Richard Michalet
Community Ecology Group
University of Bordeaux
Talence, France

Mari Moora
Institute of Ecology and Earth Sciences
University of Tartu
Tartu, Estonia

Francisco I. Pugnaire
Estación Experimental de Zonas
  Áridas
Consejo Superior de Investigaciones
  Científicas
Almería, Spain

Blaise Touzard
Community Ecology Group
University of Bordeaux
Talence, France

Martin Zobel
Institute of Ecology and Earth Sciences
University of Tartu
Tartu, Estonia

Positive Plant Interactions and Community Dynamics: An Introduction

Francisco I. Pugnaire

Life as we know it would be impossible without considering interactions between living things because each organism is related to many other creatures in one way or another. If we take a look at our lives, we see lots of various kinds of interactions with family, friends, and strangers. We compete with peers in the labor market and with other citizens for desirable goods—be these tickets for a popular concert or a table at a popular restaurant—but we also help neighbors in difficulty and receive help in turn. The first interactions could be considered negative, while the second are rather positive.

These interactions are also evident in the behavior of animals. Examples of negative impacts of one species on another are the actions of a predator catching its prey or defending its territory against a potential rival. Positive interactions and altruism characterize the conduct of social animals; we can see ants or bees in a colony working together to tend their offspring. But what happens with plants? They are sedentary organisms that never change site. We may notice simple answers to environmental clues: Plants grow better in favorable conditions and remain inactive during adverse times. Most often, direct interactions between plants go unnoticed. It can be observed, however, how some parasitic plants affect their hosts, or how vines climbing trees to catch light go as far as to deprive of light the tree they grow on. But the vast majority of plant interactions are indirect. These interactions are, however, intense and very important for the dynamics of plant communities. For example, imagine a plant that cannot grow because its neighbors, with strong root systems, are taking up most of the available water. The effect is clearly negative, and if it becomes severe, the plant less able to take up water from the soil will die. This case shows a regular process among neighboring plants: competition for resources. Another example would be a plant shading another and depriving it of such a vital resource as light. In these cases the final outcome may be the demise of the less competitive plant or species. Resources are shared by all plants, and these resources are limited in most ecosystems, so plant competition is widespread in nature.

Now imagine a very harsh and demanding environment with a dry and poor soil where only plants very tolerant to drought can grow; these are often shrubs or small trees with the ability to develop deep roots. When a shrub starts to grow, it

FIGURE 1 Facilitation plays a major role in arid ecosystems, allowing for higher species diversity. A conspicuous plant community develops under the canopy of these shrubs in the Tabernas basin in Almería, in contrast to the bare intershrub spaces.

immediately changes its surrounding zone of influence. First, by casting its shadow on the ground, it buffers daily fluctuations in temperature; then its leaf litter decomposes on the ground and increases the nutrient content of understory soil. At night, some plants may even moisten the soil surface through a process by which water extracted from deep soil layers is shed on surface layers. This process is known as hydraulic lift (Richards and Caldwell 1987). In short, plants modify the environment where they live to feel more comfortable and create more favorable conditions for growth. These new environments are used by other species that cannot tolerate the initial conditions but can establish themselves in the new site. Plants that facilitate the growth of others are termed *nurses* (Figure 1).

Over time, when the facilitated plants grow and their roots explore and obtain more soil resources, they may outgrow their nurses, which are often eliminated through competition. Nurse plants are classical examples of positive interactions between plants, without which some species and their associated fauna would not survive in some ecosystems. As evidenced by recent experiments, *facilitation*—a term used to describe positive, one-way interactions—is a very common phenomenon in plant communities and in many environments, and is not restricted only to those with extreme conditions.

Interactions between organisms have always occupied a central role in ecology. It was quite some time ago—ever since Darwin's "struggle for life" became the heart of his theory of evolution—that biologists began to appreciate the relevance of interactions for the natural history and the evolution of organisms. Modern theories also recognize the importance of this struggle, presented as competition between genes to persist and disperse as widely as possible. But the positive interactions between plants began to attract the attention of researchers only very recently, just over a dozen years ago. This interest is triggered by several factors. First, evolutionary theory is based almost exclusively on competition, without which you cannot predict or explain the most important evolutionary events; however, there may be different explanations for the development of each stage of the organizational hierarchy. These theories consider evolution, but emphasize the crucial importance of positive interactions, such as cooperation or symbiosis. In plants, it is possible to assess the importance of positive interactions because, in their absence, we could not explain the geographical distribution of many species, the invasion of exotic plants, or the potential responses of plant communities to climate change.

Concern about the weakness of current theory, which concedes to competition the leading role in ecological processes, is another cause of the recent interest in positive interactions among plants. In fact, "niche construction" by some species enhances the establishment of other species in the ecosystem, thus increasing its diversity. Higher plant diversity also means greater diversity among consumers and of other organisms that interact with plants (fungi, for example). This means a better functioning of ecosystems and, although there are still many details to understand all ecosystem processes, it is quite clear that positive interactions have a major role in supporting and maintaining biodiversity. This role is particularly important in fragile ecosystems, as is the case in dry and alpine habitats. Finally, positive interactions in general, and facilitation and nurse plants in particular, have great application potential. For example, shrubs with the ability to act as nurses can be used in ecosystem restoration, saving considerable time and effort that otherwise would require using more invasive techniques (Padilla and Pugnaire 2006).

But all plants exert positive and negative effects on their neighbors, and the balance depends on many factors, including the level of resources, the presence of herbivores, or the frequency of disturbances. What we see as the outcome of the interaction between two plants is the net balance of their interaction, i.e., the cumulative effect over time of positive and negative aspects. At any one point, positive effects may predominate over negative ones or vice versa, depending on environmental conditions (Armas and Pugnaire 2005). Although knowledge on plant interactions and their importance to ecosystem processes has greatly increased in recent years, there are still many aspects that require experimental and conceptual effort to be fully understood (Brooker et al. 2008). Chief among these are (a) the degree of nontransitivity in plant competitive networks to community diversity and facilitative promotion of species coexistence and (b) the role of indirect facilitation along environmental severity gradients.

This volume gathers a set of papers wherein authors analyze different aspects of plant interactions with an emphasis on facilitation.

Ragan M. Callaway stirred the field with his seminal paper coauthored with M. Bertness in 1994. In Chapter 1, he provides an overview of facilitation, concluding that research on facilitation explains at least in part the success of invasive plant species, and shows the existence of evolutionary relationships among plants, a decidedly nonindividualistic process, which play important roles in the organization of communities.

In Chapter 2, Kikvidze and Armas address the study of plant interactions by discussing the different approaches used over time. A number of tools are available for analyzing the direction, intensity, and importance of plant interactions and to accurately quantify them. The authors reckon, however, that this is not an easy task, as plant interactions entirely reflect the diversity, complexity, and variability of natural systems. Consequently, it is not surprising that, despite the long history of the concept of plant interactions, ecologists were not able to start experimental measurements until the last quarter of the 20th century. An array of quantitative tools reflecting many aspects of plant interactions has been developed, and efforts are continually made to improve them.

In Chapter 3, Cavieres and Badano focus on the scale at which facilitation works, assessing its effects from the individual plant to the landscape level. They show how increased species richness due to facilitation enables the coexistence of taxa with different historical and evolutionary trajectories, decreasing the chances of redundancy in ecosystem functioning. They describe how more-diverse communities may have enhanced relevant ecosystem processes (e.g., carbon and nitrogen fixation), highlighting the role of facilitation for the functioning of ecosystems. The authors show how environments in which facilitation is a key process, like alpine and arid systems, are particularly sensitive to anthropogenic effects, like climate change, land use change, and introduced species.

In Chapter 4, Michalet and Touzard analyze the role of positive and negative interactions in their relationships with biodiversity and ecosystem functioning. They use different models as well as unpublished data on riparian grasslands and old fields to reconcile diversity experiments with diversity natural patterns.

In Chapter 5, Moora and Zobel review the role of mycorrhizal symbiosis in plant–plant interactions, focusing on the effect of mycorrhizal-mediated facilitation on the structure and dynamics of plant communities. Plant facilitation and mycorrhizal symbiosis are both important processes structuring natural and anthropogenic plant communities. Mycorrhizae may influence plant facilitation either directly by changing plant traits or indirectly by influencing relationships between plants and organisms of other trophic levels. A great amount of experimental evidence supports the idea that mycorrhizae may, to a great extent, affect positive interactions between plants, having a significant effect at the whole plant community and ecosystem.

In Chapter 6, Brooker reviews the impacts of climate change on plant–plant interactions, using case studies to illustrate underlying fundamental points that are relevant to all plant communities, irrespective of their location. Evidence for such patterns and processes can be found in ecosystems all across the globe, and the concepts discussed are clear enough to be readily extrapolated beyond the confines of temperate/northern European plant communities, where most of the examples come from.

Finally, in Chapter 7, Lortie analyzes the consistency of the stress–gradient hypothesis, which predicts that the relative frequency of positive interactions between plants will increase with increasing consumer pressure or environmental stress (Bertness and Callaway 1994). He focuses on the strength of the associated logic and concepts, and discusses the results of a subset of published empirical tests of this hypothesis, explaining how gradients could be used, followed by an exploration of stress, to finally combine these ideas with an analysis of empirical tests for the stress–gradient hypothesis.

These topics comprise the bulk of research carried out in the last 10 years on plant interactions as the study of associations, interactions, and gradients has led to a tremendous growth in community ecology. Admittedly, the aspects that contribute to the outcome of plant interactions, like the plant pollinator and plant–herbivore interactions, are also major avenues of research, but for this book we choose to develop plant–plant interactions thoroughly. Pollinator and herbivore interactions need appropriate treatment separately as the primary goal of this volume is to introduce readers to the application of interaction theory. Likewise, such aspects as the use of facilitation in ecosystem management and restoration or the role of indirect interactions would also need further attention.

We need a good understanding of natural processes to be able to properly manage natural habitats, prevent environmental risks, and secure the continued supply of ecosystem services. Complex processes are at the base of ecosystem diversity, and hopefully those addressed in this volume will help us improve our understanding of nature.

REFERENCES

Armas, C., and F. I. Pugnaire. 2005. Plant interactions govern population dynamics in a semi-arid plant community. *Journal of Ecology* 93: 978–989.

Bertness, M., and R. M. Callaway. 1994. Positive interactions in communities. *Trends in Ecology and Evolution* 9: 191–193.

Brooker, R. W., F. T. Maestre, R. M. Callaway, C. L. Lortie, L. A. Cavieres, G. Kunstler, P. Liancourt, K. Tielbörger, J. M. J. Travis, F. Anthelme, C. Armas, L. Coll, E. Corcket, S. Delzon, E. Forey, Z. Kikvidze, J. Olofsson, F. I. Pugnaire, C. L. Quiroz, P. Saccone, K. Schiffers, M. Seifan, B. Touzard, and R. Michalet. 2008. Facilitation in plant communities: The past, the present and the future. *Journal of Ecology* 96: 18–34.

Padilla, F. M., and F. I. Pugnaire. 2006. The role of nurse plants in the restoration of degraded environments. *Frontiers in Ecology and the Environment* 4: 196–202.

Richards, J. H., and M. M. Caldwell. 1987. Hydraulic lift: Substantial nocturnal water transport between soil layers by *Artemisia tridentata* roots. *Oecologia* 73: 486–489.

1 Do Positive Interactions among Plants Matter?

Ragan M. Callaway

CONTENTS

1.1 INTRODUCTION

Our perception of nature is based, to a large degree, on observing how organisms interact with one another. For example, if our observations are predominantly of consumer relationships, then we are likely to lean strongly toward a "red in tooth and claw" perspective. If we tend to focus on mutualistic and beneficial interactions, then we may be more likely to perceive the natural world as Gaia. Alternatively, of course, our perception of nature may determine what interactions we find interesting or important. For example, one who inherently sees the world as red in tooth and claw may be predisposed to consider predation to be a dominant interaction in nature.

Fundamental theory about the nature of ecological communities has also probably been affected by perspectives on how species interact, and, vice versa, perspectives on the nature of communities have probably affected our tendency to prioritize the importance of different kinds of interactions among species. For example, for over 50 years most plant ecologists have accepted that the distribution of plant species, and their organization into groups or communities, is determined individualistically, that is, by the adaptation of each species in a "community" to a particular abiotic environment, highly stochastic dispersal events, competition among these similarly adapted species, and the disruption of adaptive and competitive distributions by consumers and disturbance. Definitions of the "individualistic" paradigm of plant community organization are not identical (see Nicolson and McIntosh 2002), but in general they emphasize "the fluctuating and fortuitous immigration of plants and an equally fluctuating and variable environment" (Gleason 1926). Moore (1990, see also Nicolson and McIntosh 2002) rephrased the individualistic concept as "vegetation as an assembly of individual plants belonging to different species distributed according to its own physiological requirements as constrained by competitive interactions."

1

The individualistic view of plant communities has led to highly successful research on the abiotic environment and competition as factors structuring plant communities. Negative interactions such as predation, competition for resources, and allelopathy are central to the study of ecology and evolution. However, it is becoming increasingly clear that many organisms, and in particular plants, can greatly improve the growth and fitness of their neighbors. Positive interactions among plants, or *facilitation*, occur when the presence of one plant enhances the growth, survival, or reproduction of a neighbor. Much like the way the term *competition* is used in the literature, the term *facilitation* is also used in a loose manner, and facilitation may occur with negative, positive, or neutral reciprocal responses from neighbors. The goal of this chapter is to consider whether or not these facilitative interactions really matter in the larger context of community ecology.

Any definition of facilitation suggests something fundamentally different from the individualistic paradigm. If the presence of one plant can *increase* another species' fitness or the probability that another species may occur in the same place, plant communities cannot be individualistic. In the last 20 years, hundreds of peer-reviewed papers have been published on the topic of facilitation (Callaway 2007). This research challenges a strict definition of individualistic plant communities, one of the most basic and widely accepted conceptual models in ecology, as a foundation for understanding how groups of plant species are organized. The implications of rethinking plant individualism go beyond theory; if plant communities are less individualistic than we have thought, the conservation implications are profound (see Padilla and Pugnaire 2006). Clearly, positive interactions matter to general conceptual community theory; they demonstrate the inadequacy of the venerable individualistic paradigm. However, burgeoning research on facilitation has highlighted other ways in which positive interactions may matter.

1.2 FACILITATION AND PRODUCTIVITY

When overstory trees increase the productivity of the plants growing beneath them, the combination of tree productivity and herbaceous productivity is obviously much higher than in open areas without trees. For example, in woodlands in central California, the facilitative effects of *Quercus douglasii* on herbaceous productivity can increase grassland productivity by two to three times that in the open, and total tree leaf and stem production of the oaks is estimated to be even greater than that of the herbs (Holland 1973; Callaway et al. 1991; Ratliff et al. 1991; R. M. Callaway, unpublished data). Thus the beneficial effects of a single species can dramatically alter energy flow and storage in natural ecosystems.

Many other studies have shown increased biomass through facilitation (e.g., Barth 1978; Rumbaugh, Johnson, and Van Epps 1982; Jackson and Ash 1998; Thomas and Bowman 1998; Schade et al. 2003; Zhang and Li 2003), but one of the most thorough investigations of facilitation on ecosystem productivity was conducted by Patten (1978) in the Sonoran Desert, which wraps around the Gulf of California in North America. He harvested ephemerals in habitats created by trees and shrubs and in open spaces throughout two growing seasons and dried the harvested material for mass and caloric measurements. Mass measurements allowed computation of standard final biomass, and repeated measurements of mass allowed computation of rates of productivity. Caloric measurements were compared to solar input to determine production efficiencies (percent solar energy used). Directly beneath the small tree, *Cercidium microphyllum*, seasonal rates of primary productivity (SPP) were 1.45 to 1.62 kg^{-1}ha^{-1} day^{-1}, and under the shrub, *Ambrosia dumosa*, SPP reached 0.48 to 0.42 kg^{-1}ha^{-1}day^{-1} in a wet year. In the same year, SPP rates at the edge of *Cercidium* canopies were 0.76 and 0.37 kg^{-1}ha^{-1}day^{-1}, and in the open the rate was 0.36 kg^{-1}ha^{-1}day^{-1}. Although total productivity was much lower in a subsequent dry year, even stronger facilitation patterns were apparent for annuals under *Cercidium* (0.418 beneath to 0.045 in open).

Other interesting patterns were clear when productivity was converted to percent production efficiency (total cumulative solar energy entering a habitat divided by the caloric content of the annuals). The effect of *Cercidium* was even more striking when these calculations were made. In the open areas without perennial canopies, production efficiency was 0.02% of solar energy converted to caloric energy. Under *Cercidium* this was over 30 times higher, 0.68%.

To my knowledge, Patten's study is the only one in which facilitative effects have been calculated as production (biomass), productivity (kg^{-1}ha^{-1}day^{-1}), and production efficiency (chemical energy stored as a function of solar energy input). Quantified facilitative effects on productivity were +40% calculated as biomass, +830% calculated as productivity, and +3,300% calculated as production efficiency because substantially less light energy was available under canopies. Patten measured the growth of ephemeral species that completed their life spans in weeks or months. Virtually all other similar studies reasonably assume that biomass of annual species at the end of the growing season is a reliable measure of annual productivity. Patten's results indicate that there is a much greater turnover of species or far more recruitment throughout the year in the understory of *Cercidium* than in the open; otherwise, *Cercidium*'s effects on biomass would be much more similar to its effects on productivity. If biomass and productivity are this disconnected in other ecosystems, we often may be substantially underestimating facilitative effects. The choice to use biomass or productivity as a metric for the effects of *Ambrosia* would appear to be even more important. The two-year average of biomass under *Ambrosia* indicated a 72% decrease relative to the open. However, comparing Patten's calculations of seasonal productivity indicated a 20% *increase* under *Ambrosia*.

Early successional *Pinus albicaulis* (white bark pine) trees appear to facilitate *Abies lasiocarpa* (subalpine fir) near the timberline under some environmental conditions in the Northern Rocky Mountains (Callaway 1998). In this context, Callaway, Sala, and Keane (2000) quantified the biomass allocation and productivity of harvested individual *P. albicaulis* trees and used these age-allocation relationships to estimate productivity in natural forests by applying regression models for harvested trees to all trees in each stand. Stands varied from 67 to 458 years in age. They hypothesized that that the increasing (and potentially facilitated) abundance of *A. lasiocarpa* during succession might maintain high rates of annual productivity as these forests aged. The estimated productivity of *P. albicaulis* in these forests increased for approximately 200–300 years and then slowly decreased over the next 200 years. In contrast, as stands shifted in dominance from pine to *A. lasiocarpa* with age, *A. lasiocarpa* appeared to maintain gradually increasing rates of whole-forest productivity until stands were approximately 400 years old. Although evidence for the facilitative effects of *P. albicaulis* and facilitated responses of *A. lasiocarpa* is only for high-elevation timberline forests, it suggests that facilitative processes might play an important role in maintaining high rates of productivity in some sub-alpine forests.

1.3 POSITIVE INTERACTIONS AND DIVERSITY–ECOSYSTEM FUNCTION

The effects of diversity on ecosystem functioning and emergent community properties has been a primary focus in modern ecology (Chapin et al. 2000; Loreau et al. 2001; Hooper et al. 2005). A great deal of experimental research, based primarily on grassland communities, has found relationships between increasing plant diversity and ecosystem attributes (Tilman, Wedin, and Knops 1996; Naeem et al. 1996; Tilman et al. 1997, 2001; Hooper and Vitousek 1997; Hector et al. 1999; Hooper et al. 2005). Clearly, direct and indirect positive effects of species on one another have the potential to drive the relationship between community diversity and ecosystem productivity, stability, invasibility, and resource cycling. However, despite the ubiquitous occurrence of facilitation in natural communities, positive interactions have not received much attention in research on diversity–ecosystem function. Most ecologists attribute the relationship between increasing diversity and functional attributes to *complementarity*. Complementarity occurs when performance of a species is greater because neighbors do not substantially infringe on its resource requirements, i.e., reduced competition. In contrast, facilitation is when a species actually benefits from the presence of its neighbor. Separating these very different conceptual mechanisms may allow a better understanding of the role of diversity in community and ecosystem functioning.

An important issue in diversity–ecosystem function research is determining the relative importance of diversity per se, versus the effects of particular species on ecosystem function (Spehn et al. 2002). Facilitation is often caused by particular traits, such as the ability to hydraulically lift water, mine nutrients from deep soils, bind mobile substrate, or tolerate high winds, suggesting that species-specific effects should be important. However, facilitation can also be caused by nonspecific processes such as whole-community boundary layer effects (Mulder, Uliassi, and Doak 2001), and such effects may contribute to the role of diversity per se on ecosystem functioning. Indirect facilitation involving consumers, if due to "shared resistance," is not likely to affect ecosystem function through diversity per se. This is because shared resistance is derived from particular defense traits of species. In contrast, indirect facilitation due to "associational resistance" is likely to drive diversity–function relationships through the effects of diversity because associational resistance is derived simply through the complexity of community structure and chemical composition.

Facilitation caused by indirect interactions among competitors may be an overlooked process in the diversity–ecosystem function debate. This is because indirect interactions among competitors are extremely difficult to identify, much less tie experimentally to ecosystem function. Indirect interaction among competitors may be species-specific in some cases, and when so, it is likely to drive diversity–function relationships via the presence of certain species. However, the phenomenally complex indirect webs of interactions that may occur among competing plants (Miller 1994) are likely to affect ecosystem function through the diversity per se of a community. To my knowledge, the role of indirect facilitation among competitors has never been explored as a potential cause of diversity–function relationships.

One of the best experimental analyses of a facilitative mechanism behind the effect of diversity on function has been carried out on stream invertebrates. Cardinale, Palmer, and Collins (2002) showed that experimentally increasing the species diversity of stream invertebrates induced facilitative interactions and led to nonadditive changes in resource consumption. This occurred because increasing diversity reduced the deceleration of water flow from upstream to downstream neighbors, allowing more diverse communities to acquire a higher proportion of suspended resources than were captured by monocultures. This difference was based on fundamental differences in the thermodynamic properties of diverse communities, which raises the interesting possibility that similar processes may occur in terrestrial systems. In terrestrial systems "engineered," or facilitated, complexity and diversity can alter fluxes of energy and resources (Jones, Lawton, and Shachak 1994, 1997). As stated by Cardinale, "Changes in species diversity may alter the probability of positive species interactions, resulting in disproportionally large changes in the functioning of ecosystems."

Strong facilitative interactions must increase community diversity, especially in cases in which beneficiaries have an obligatory or near-obligatory dependence on benefactors. However, whether or not increased diversity, driven by facilitation, feeds back to ecosystem function is less certain. Many studies have found that the relative importance of facilitation increases with abiotic stress (e.g., Greenlee and Callaway 1996; Callaway 1998; Choler, Michalet, and Callaway 2001; Pugnaire and Luque 2001). Mulder, Uliassi, and Doak (2001) reasoned that if environmental stress creates conditions in which facilitation is important, then environmental stress may also create conditions in which plant species richness increases community productivity. In other words, "Species that seem to be functionally redundant under constant conditions may add to community functioning under variable conditions" (Mulder, Uliassi, and Doak 2001). They established experimental communities of bryophytes found growing near each other in New Zealand. In control conditions, where humidity was high and light intensity was low, there was no relationship between species richness and community productivity. However, in the bryophyte communities exposed to short-term drought, the biomass of all species combined increased with the species richness of the community. Mulder et al. explored two different hypotheses to explain their results: (a) the insurance hypothesis, which posits that increased productivity in drought is due to the increased probability that the community contains species that are capable of dominating under the changed environmental conditions, and (b) the positive interactions hypothesis, wherein an increase in positive interactions in drought conditions among plants drives the relationship between diversity and productivity. They found that increased biomass under drought conditions was associated with increased survivorship for most species. Biomass per surviving plant did not change significantly with species richness in drought, but species demonstrating the greatest increases in biomass in the drought conditions were the least resistant to drought. Finally, they found that for a given species pool, the number of species contributing positively to biomass increases also increased with diversity. Mulder et al. argued that an increase in positive interactions among plants in stress conditions was the reason for increased productivity with species diversity. These positive interactions could have occurred due to a general increase in humidity, with plants with different architectures creating a more complex community boundary layer in diverse communities and trapping more transpiration, or taller bryophytes may have protected shorter species from photoinhibition, a decrease in photosynthetic capacity that often increases in drought conditions.

In Portugal, Caldeira et al. (2001) constructed plots with different numbers of Mediterranean grasses and forbs to investigate mechanisms that might drive diversity–productivity relationships. After two years they measured plant cover, the number of species present, soil moisture in their plots, and stable carbon isotope ratios ($\delta^{13}C$) in the leaves of five different species with C_3 physiology. They found that the total biomass and total cover in species-rich plots were significantly higher than in monocultures. However, their findings did not indicate niche complementarity as the driving mechanism. The niche complementarity hypothesis would be supported if mixtures of species performed better as a *community*, but not as *individuals*. In other words, enhanced community performance would simply be due to the inclusion of species that were each better at exploiting resources in a particular dimension of space or time. Caldeira and her colleagues found that *individuals* performed better in mixtures. For four of the five individual species tested, leaf $\delta^{13}C$ indicated significantly higher water-use efficiencies in the species-rich mixtures than in monocultures. CO_2 gradients were minimal, leaf temperatures varied less than 1°C, and there were no differences in leaf nitrogen concentration, suggesting that differences in $\delta^{13}C$ were due to environmental modification within diverse communities and not to differences in photosynthetic capacity among species. Niche complementarity also may have occurred, of course (suggested by lower soil water content in plots with mixtures); however, this study, and that of Mulder, Uliassi, and Doak (2001), opens a window into a new way of thinking about facilitation and community productivity.

Interspecific attraction of pollinators or dispersers, shared defenses against herbivores, or resistance to pathogens may all increase community diversity, and the indirect facilitative interactions caused by these organisms can drive the effects of community diversity on ecosystem function. Knops et al. (1999) built experimental communities that varied in species diversity and assessed disease severity. For each of four target plant species, foliar disease was significantly negatively correlated with plant species richness. However, they found that disease severity was more strongly dependent on host plant density and not on richness per se, suggesting that disease transmission may have been simply a function of the richer plots having lower densities of host plants.

Spehn et al. (2002) investigated the effects of plant diversity on nitrogen accumulation in plants. They varied the number of plant species and functional groups (grasses, herbs, and legumes) in experimental grassland communities across seven

European experimental sites, and two years after starting the experiment they found that diversity and community composition had strong effects on ecosystem properties. Two years after sowing, nitrogen pools in Germany and Switzerland strongly increased in the presence of legumes, but less so at other sites. Over all sites, nitrogen concentration was very similar regardless of the presence of legumes, averaging $1.66 \pm 0.03\%$ across all sites and diversity treatments. But legumes had a positive effect on nitrogen by significantly increasing aboveground plant biomass. The legume effect was the strongest at the German site, and there nitrogen fixed by legumes was transferred to grasses and forbs. There was a high degree of species-specificity in this process, and transfer depended on the particular legume species fixing N and the particular grass or forb taking it up. These results demonstrate that the facilitative effects caused by nutrient addition may contribute to the effects of diversity on ecosystem function.

If the causal relationship between facilitation and productivity indicated by Spehn, Caldeira, and Mulder's research is a broad phenomenon, understanding this relationship has the potential to profoundly alter the way we understand the role of positive interactions in community structure. The widespread idea that some species may be "redundant" in communities is not acceptable until so-called redundancy is shown in a realistically broad range of environmental conditions experienced by a community in nature. Ignoring the possibility that "redundant" species in nonstressful conditions may either elicit or respond to facilitative mechanisms in stressful conditions will result in an underestimation of the value of species diversity in plant communities.

Interestingly, other theoretical research indicates that positive interactions may produce species-rich communities (Gross 2008). Gross showed that important mechanisms for coexistence emerged in models that explicitly incorporated simultaneously occurring negative and positive interactions. Likewise, in a phylogenetic analysis of a global database, Valiente-Banuet and Verdu (2007) found that the use of regeneration niches constructed by facilitators was strongly conserved across evolutionary history and, by predominantly facilitating more distantly related taxa, nurse species increased phylogenetic diversity.

1.4   FACILITATION AND EXOTIC INVASION

Despite a growing body of research on how exotic and native plants interact and a growing body of research on facilitation among plants, we do not know much yet about the importance of facilitation in exotic invasion (see Bulleri, Bruno, and Beneditti-Cecchi 2008). However, several studies indicate that positive interactions may matter for some plant invasions.

Freeman and Emlen (1995) inferred interactive strengths from spatial associations and from the performance of many species in a shrub–steppe community in western Utah. They found that the invader *Bromus tectorum* (cheatgrass) did not appear to be negatively affected by any of the native species present, a pattern not true for any of the native species. In fact, native species "actually facilitated the reproduction of *Bromus tectorum*." It certainly does not appear that *B. tectorum* required facilitation to invade, but facilitation may have augmented the considerable invasive potential of this species.

Lupinus arboreus (bush lupine) is native to California and derives 60%–70% of its nitrogen budget from nitrogen fixation (Bently and Johnson 1991). *Lupinus arboreus* may add as much as 185 kg/ha of nitrogen to subcanopy soils (Gadgil 1971; Palaniappan, Mars, and Bradshaw 1979). Soil under *L. arboreus* shrubs also has approximately twice the exchangeable ammonium concentration and three to five times the exchangeable nitrate concentration as soils away from shrubs (Maron and Connors 1996). They found that the total aboveground biomass of herbaceous species was more than twice as much in dead *Lupinus* patches as in lupine-free grassland. Additionally, the annual grass *Bromus diandrus* grew much larger in soil collected from *Lupinus* patches compared to soil from the open grassland. However, this facilitative effect was manifest primarily for nonnative species such as *B. diandrus*, and nitrogen-rich patches associated with dead *Lupinus* shrubs were focal points for the facilitation and establishment of many exotic, weedy species. Dead *Lupinus* patches contained 47% fewer plant species overall and 57% fewer native species. Other experiments conducted by Pickart, Miller, and Duebendorfer (1998) showed that the removal of lupine alone roughly doubled the cover of native species and had minimal effects on the cover of exotic species. However, removal of the lupine plant and the duff layer increased the percent cover of native species ≈5× and decreased the cover of exotic grasses from 35% to 0%. They also concluded that the nonnative species played a role in maintaining enriched soil nitrogen. Maron and Jefferies (1999) found that 57%–70% of the net amount of nitrogen mineralized annually was taken up by the annual grasses that occupied dead lupine patches and returned to the soil when the annuals died. They estimated that the facilitative effect of the lupines on soil fertility would take 25 years to decline 50%, which demonstrates the longevity of some positive effects. Similar patterns were reported by Carino and Daehler (2002), who found that an inconspicuous annual legume, *Chamaecrista nictitans*, facilitated the invasion of *Pennisetum setaceum* (fountain grass) into native *Heteropogon contortus* grasslands in Hawaii.

Over 60 years ago, Went (1942) observed that the positive effect of shrub species on annuals tended to be more evident after benefactor mortality than before the facilitator died. Now, in the region where Went worked, facilitation of natives by shrubs has been replaced to a large degree by facilitation of exotic invaders. Claus Holzapfel and Bruce Mahall (1999) found that native shrubs clearly facilitated exotic grasses and herbs. Considering that several of the invasive annuals they worked with occur at much lower abundances in their native land, it is possible that alien plants have altered the natural balances among the types of mechanisms that existed among native shrubs and herbs.

Much like the patterns that emerged from Holzapfel and Mahall's (1999) research, the positive effects of *Quercus douglasii* on understory productivity described at the beginning of this chapter was manifest almost exclusively through the facilitation of exotic Eurasian grasses such as *Bromus diandrus* (Callaway, Nadkarni, and Mahall 1991). Facilitative effects did not occur for the sole remaining native bunchgrass species, *Nassella* (*Stipa*) *pulchra*.

Siemann and Rogers (2003) observed that the first *Sapium sebiferum* (Chinese tallow) trees invading grasslands of southern Texas appeared to create high nitrogen and low light conditions that favored their offspring seedlings in competition with the native herbaceous vegetation. They mimicked the effect of *Sapium* by manipulating nitrogen and light in field experiments. The growth of *Sapium* seedlings increased with nitrogen fertilization. Under shading, the aboveground biomass of prairie vegetation did not change; prairie vegetation biomass decreased; and tree seedling growth increased. *Sapium* growth increased dramatically in the treatment with both nitrogen and shade. However, in a second experiment, *Sapium* growth increased in higher light, suggesting that greater *Sapium* growth at low light levels in the first experiment was probably a consequence of decreased competitive interference from prairie vegetation, rather than the preference of *Sapium* for low light. Although they did not manipulate the effects of mature trees on seedlings, their results suggest that intraspecific facilitation, perhaps the "invasional meltdown" of Simberloff and von Holle (1999), are involved in *Sapium* invasions of grasslands.

Chenopod shrublands in southern Australia are being invaded by the succulent Asclepiadaceae, *Orbea variegata*. Lenz and Facelli (2003) observed that *Orbea* often grows beneath native shrubs. They conducted a suite of experiments designed to ascertain whether or not facilitation was important to the invasion, and if so, what mechanisms were involved. Their results indicated that reduction in light and temperature under shrubs, rather than increased nutrients in subcanopy soils, are the primary facilitative mechanisms that benefit *Orbea*. Temperatures above 30°C, which are more likely to occur on the soil surface of open areas than under shrubs, inhibited seed germination. Seedling survival and the growth of established ramets were increased by shade. They concluded that the facilitative effect of shrubs was not obligatory, which is common in facilitative relationships in general, but that *Orbea* gained substantially from the presence of the shrubs.

Nonnative invasive species can modify vegetation establishment after disturbance. *Myrica faya*, an invasive nitrogen-fixer, establishes after volcanic disturbance and strongly inhibits the establishment and growth of native species (Vitousek et al. 1987; Vitousek and Walker 1989; Walker and Vitousek 1991). On the volcanic Mount Koma in Japan, native communities under nonnative *Larix kaempferi* canopies have significantly greater richness and diversity than communities under the native *Betula ermanii* or in the open (Titus and Tsuyuzaki 2003). They hypothesized that *Larix* accelerates succession of a few smaller native species, but that overall succession will be deflected toward dominance by the introduced species.

Heather Davis and colleagues (2004) at UC Davis documented strong Allee effects for invading *Spartina alterniflora*. Early invading, isolated *S. alterniflora* plants set about one-tenth the seed as plants growing in denser stands, but as these recruits increase in density as plants grow clonally, they produce more viable seeds.

In summary, the growing body of evidence for the importance of positive interactions in exotic plant invasions points to a crucial yet relatively unexplored research area.

1.5   FACILITATION AND EVOLUTION IN PLANT COMMUNITIES

Facilitative relationships among plants certainly do not imply evolutionary relationships. However, by definition, plant species interacting in individualistic communities cannot exert selective forces on one another. With few exceptions (Aarssen, Turkington, and Cavers 1979; Turkington 1989; Menchaca and Connolly 1990; Turkington and

Mehrhoff 1990), plant communities are not thought to consist of co-evolved species, or to possess stable inherent properties determined by plant–plant interactions.

Despite the predominance of the individualistic paradigm, recent research raises the possibility of evolutionary relationships within plant communities and the potential of facilitative interactions to play a role. Figueroa et al. (2003) examined invasion of the Odiel Marshes in the joint estuary of the Odiel and Tinto Rivers in southwest Spain by a *Sarcocornia* species and found the invader to be a hybrid. The *Sarcocornia* hybrid appears to benefit from an unusual successional sequence in which the raised centers of *Spartina martima* patches are invaded by *Sarcocornia perennis*, a species common to lower marshes. However, once established in *Spartina* patches, *Sarcocornia perennis* provides an opportunity for hybridization with *Sarcocornia fruticosa*, a species common in the higher marshes. Hybrids only occur on *Spartina* patches with *Sarcocornia perennis*. Figueroa et al. coined the establishment of the hybrid *Sarcocornia* as *genetic facilitation*, and suggested that succession might be facilitated genetically through the establishment of conditions leading to hybridization rather than simply by the enhanced sediment accretion by earlier species. Sediment accretion prevents later successional marsh species from being submersed, and prevents anaerobic conditions in the low marsh that appear to exclude *Sarcocornia fruticosa*.

As described in detail here and elsewhere (Callaway 2007), plants have powerful modifying forces on their environments. This effect has been coined as *ecosystem engineering* by Clive Jones and colleagues, although engineering is not limited to plants. Plants can favorably alter the availability of all fundamental resources (nutrients, water, and light), change the way that energy and materials are cycled, and profoundly alter the course of natural disturbance.

Theoretical models have been produced in which ecosystem engineering, also called *niche construction*, is responsible for selective feedback in evolution (Laland, Odling-Smee, and Feldman 1996, 1999; Odling-Smee, Laland, and Feldman 1996). In these models niche construction by early generations of a species can affect the evolution of their offspring. In other words, the developers of these models argue that the effects of plants on patterns of light and shade, soil salinity, and nutrient cycling have the potential to alter future evolutionary trajectories. The first model developed by Laland, Odling-Smee, and Feldman (1996) assumed that the abundance of a key resource depended entirely on niche construction effects. With this constraint, they found that niche construction could cause either evolutionary inertia or momentum, cause the fixation of otherwise deleterious alleles, establish unexpected stable polymorphisms, or eliminate expected stable polymorphisms. Clearly, the availability of resources is determined by environmental modification, and in a second model both niche construction and independent abiotic processes were included (Laland, Odling-Smee, and Feldman 1999). The results of the second model supported the conclusion of the first. Even with independent, abiotic effects on resource availability, the effects of environmental modification by an organism had the potential to override external sources of genetic selection to "create new evolutionary trajectories and equilibria, generate and eliminate polymorphisms, and produce time lags in the response to selection as well as other unusual dynamics." When the effects of environmental modification, or niche construction, opposed the external source of selection, they were particularly strong.

Most facilitative effects described in the literature almost always oppose the selective effects of the abiotic environment. It has been exceptionally difficult to integrate ecosystem and community ecology with the evolutionary progress of population ecology because the abiotic components of ecosystems do not evolve; however, models such as the one developed by Laland, Odling-Smee, and Feldman are crucial steps toward linking these fields and understanding the full evolutionary potential of facilitative interactions.

Atsatt and O'Dowd (1976), in their landmark paper on plant "defense guilds" and herbivore-driven functional interdependence among plants, argued that groups of "populations, races, closely related species, and unrelated but chemically similar species" form "gene conservation guilds." In short, by forming guilds with other species, plants may make it far more difficult for pathogens, parasites, and herbivores to win the genetic "arms race." This arms race has made it exceptionally difficult to sustain agricultural monocultures free of disease, herbivores, and pathogens without intense applications of pesticides and biological controls. In part, this is due to the ease with which pathogens and herbivores can find their prey and multiply rapidly, but monoculture susceptibility is also due in part to substantial disadvantages for the consumed relative to the consumer in the evolutionary race between offense and defense.

Genetic uniformity, whether within or among species, provides unique opportunities for "evolutionary tracking"—the evolution of virulent pathogens and herbivores, which can be rapidly selected for by monocultural hosts. Heterogeneity in host populations or communities decreases the exposure frequency of susceptible genotypes and can disrupt evolutionary tracking and specialization by pathogens and herbivores. If pathogens and herbivores cannot specialize, they may be much less likely to become virulent. To summarize, in heterogeneous mixtures of plants, consumers may have to spend some of their time eating plant species that select for different consumer characteristics than would other plant species. When this occurs, the consumer cannot be maximally selected for either species and therefore does not become exceptionally virulent against either species. Most of the best evidence for these sorts of "gene conservation guilds" is from agriculture and focuses on different genotypes of crop species. However, there is a great deal of potential for gene conservation guilds to function as powerful facilitative mechanisms in natural plant communities, and these guilds may contribute to a greater understanding of community diversity and function.

If the sort of decoy–target effect on herbivores demonstrated in the potato–potato beetle system described in Atsatt and O'Dowd is controlled in any way by the relative abundance of neighbors, the interactions may vary in function over time and space. For generalist herbivores, the amount of an unpalatable species that is eaten depends somewhat on how abundant palatable species are in the same area. For example, when only toxic phenotypes of *Lotus corniculatus* are available, the slug *Agriolimax reticulatus* consumes it. However, when nontoxic phenotypes are mixed with toxic phenotypes, *Agriolimax* avoids the toxic phenotypes that it so readily eats when there are no other choices. If one studied the relationship between *Agriolimax* and toxic phenotypes without considering the effect of the nontoxic phenotypes, selective pressures on these species could be completely misunderstood.

Bodil Ehlers and colleagues (Ehlers and Thompson 2004; Grøndahl and Ehlers 2008) have studied the interplay between facilitation of herbaceous species by *Thymus* species and the allelopathic effects that different chemotypes of *Thymus* have on the herbs. All *Thymus* species produce phenolic monoterpenes that modify the soil through litter decomposition and leaching of water-soluble compounds from leaves. However, some *Thymus* species, and some chemotypes of *T. vulgaris*, also produce nonphenolic monoterpenes. *Bromus erectus* seeds were collected from plants growing in close association with *T. vulgaris* from three "phenolic" populations and three "nonphenolic populations." Seeds and seedlings were then grown in soils from all sites in a reciprocal transplant experiment. *Bromus* originally from soils beneath nonphenolic chemotypes performed significantly better on their home soil than on soil beneath phenolic chemotypes. This response to local chemotypes was observed only for soil collected directly underneath *Thymus* plants and not for soil collected near *Thymus* plants, but away from the effect of *Thymus* canopies. These results suggest that nonphenolic patches of *Thymus* may represent evolutionary hotspots (e.g., Thompson 2005) for *Bromus*. Similar results were reported for different species of *Thymus* and neighboring species (Grøndahl and Ehlers 2008).

Kikvidze and Callaway (2009) recently proposed generalizations about how facilitation integrates into evolutionary theory. They argued that natural selection based on competition provides a complete theory for speciation, but an incomplete theory for major evolutionary transitions such as the emergence of cells, organisms, and eusocial populations. They also proposed that the successful theories that have been developed for these major transitions consistently focus on positive interactions.

In sum, research on facilitation supports other work on invasive plants (Callaway et al. 2002; Vivanco et al. 2004; Callaway et al. 2005). These results indicate that evolutionary relationships among plants, a decidedly nonindividualistic process, may play important roles in the organization of communities.

REFERENCES

Aarssen, L. W., R. Turkington, and P. B. Cavers. 1979. Neighbour relationships in grass/legume communities, II: Temporal stability and community evolution. *Canadian Journal of Botany* 57: 2695–2703.

Atsatt, P. R., and D. O'Dowd. 1976. Plant defense guilds. *Science* 193: 24–29.

Barth, R. C., and J. O. Klemmedson. 1978. Shrub-induced spatial patterns of dry matter, nitrogen, and organic carbon. *Soil Science Society of America* 42: 804–809.

Bently, B. L., and N. D. Johnson. 1991. Plants as food for herbivores: The roles of nitrogen fixation and carbon dioxide enrichment. In *Plant animal interactions: Evolutionary ecology in tropical and temperate regions,* ed. P. W. Price, T. M. Lewinsohn, G. W. Fernandez, and W. W. Benson, 257–272. New York: Wiley.

Bulleri, F., J. F. Bruno, and L. Benedetti-Cecchi. 2008. Beyond competition: Incorporating positive interactions between species to predict ecosystem invasibility. *PLoS Biology* 6 (6): e162. http://www.plosbiology.org/article/info:doi/10.1371/journal.pbio.0060162.

Caldeira, M. C., R. J. Ryel, J. H. Lawton, and J. S. Pereira. 2001. Mechanisms of positive biodiversity-production relationships: Insights provided by 13C analysis in experimental Mediterranean grassland plots. *Ecology Letters* 4: 439–443.

Callaway, R. M. 1998. Competition and facilitation on elevation gradients in subalpine forests of the northern Rocky Mountains, USA. *Oikos* 82: 561–573.

Callaway, R. M. 2007. *Positive interactions and interdependence in plant communities.* Dordrecht, the Netherlands: Springer.

Callaway, R. M., R. W. Brooker, P. Choler, Z. Kikvidze, C. J. Lortie, R. Michalet, L. Paolini, et al. 2002. Positive interactions among alpine plants increase with stress. *Nature* 417: 844–848.

Callaway, R. M., N. M. Nadkarni, and B. E. Mahall. 1991. Facilitation and interference of *Quercus douglasii* on understory productivity in central California. *Ecology* 72: 1484–1499.

Callaway, R. M., W. M. Ridenour, T. Laboski, T. Weir, and J. M. Vivanco. 2005. Natural selection for resistance to the allelopathic effects of invasive plants. *Journal of Ecology* 93: 576–583.

Callaway, R. M., A. Sala, and R. Keane. 2000. Succession may maintain high leaf area:sapwood ratios and productivity in old subalpine forests. *Ecosystems* 3: 254–268.

Cardinale, B. J., M. A. Palmer, and S. L. Collins. 2002. Species diversity enhances ecosystem functioning through interspecific facilitation. *Nature* 415: 426–429.

Carino, D. A., and C. C. Daehler. 2002. Can inconspicuous legumes facilitate alien grass invasions? Partridge peas and fountain grass in Hawai'i. *Ecography* 25: 33–41.

Chapin, F. S. III., E. S. Zavaleta, V. T. Eviner, R. L. Naylor, P. M. Vitousek, H. L. Reynolds, D. U. Hooper, et al. 2000. Consequences of changing biodiversity. *Nature* 405: 234–242.

Choler, P., R. Michalet, and R. M. Callaway. 2001. Facilitation and competition on gradients in alpine plant communities: Revisiting the "individualistic" hypothesis. *Ecology* 82: 3295–3308.

Davis, H. G., C. M. Taylor, J. C. Civille, and D. R. Strong. 2004. An Allee effect at the front of a plant invasion: *Spartina* in a Pacific estuary. *Journal of Ecology* 92: 321–327.

Ehlers, B. K., and J. Thompson. 2004. Do co-occurring plant species adapt to one another? The response of *Bromus erectus* to the presence of different *Thymus vulgaris* chemotypes. *Oecologia* 141: 511–518.

Figueroa, M. E., J. M. Castillo, S. Redondo, T. Luque, E. M. Castellanos, F. J. Nieva, C. J. Luque, A. E. Rubio-Casal, and A. J. Davy. 2003. Facilitated invasion by hybridization of *Sarcocornia* species in a salt-marsh succession. *Journal of Ecology* 91: 616–626.

Freeman, D. C., and J. M. Emlen. 1995. Assessment of interspecific interactions in plant communities: An illustration from the cold desert saltbush grasslands of North America. *Journal of Arid Environments* 31: 179–198.

Gadgil, R. L. 1971. The nutritional role of *Lupinus arboreus* in coastal sand dune forestry: 3, nitrogen distribution in the ecosystem before planting. *Plant and Soil* 35: 113–126.

Gleason, H. A. 1926. The individualistic concept of the plant association. *Bulletin of the Torrey Botanical Club* 53: 7–27.

Greenlee, J. T., and R. M. Callaway. 1996. Abiotic stress and the relative importance of interference and facilitation in montane bunchgrass communities in western Montana. *American Naturalist* 148: 386–396.

Grøndahl, E., and B. K. Ehlers. 2008. Local adaptation to biotic factors: Reciprocal transplants of four species associated with aromatic *Thymus pulegioides* and *T. serpyllum. Journal of Ecology* 96: 981–992.

Gross, K. 2008. Positive interactions among competitors can produce species-rich communities. *Ecology Letters* 11: 929–936.

Hector, A., B. Schmid, C. Beierkuhnlein, M. C. Caldeira, M. Diemer, P. G. Dimitrakopoulos, J. A. Finn, et al. 1999. Plant diversity and productivity experiments in European grasslands. *Science* 286: 1123–1127.

Holland, V. L. 1973. A study of soil and vegetation under *Quercus douglasii* compared to open grassland. Dissertation, University of California, Berkeley.

Holzapfel, C., and B. E. Mahall. 1999. Bi-directional facilitation and interference between shrubs and associated annuals in the Mojave Desert. *Ecology* 80: 1747–1761.

Hooper, D. U., F. S. Chapin III, J. J. Ewel, A. Hector, P. Inchausti, S. Lavorel, J. H. Lawton, et al. 2005. Effects of biodiversity on ecosystem functioning: A consensus of current knowledge. *Ecological Monographs* 75: 3–35.

Jackson, J., and A. J. Ash. 1998. Tree-grass relationships in open eucalypt woodlands of northeastern Australia: Influence of trees on pasture production, forage quality and species diversity. *Agroforestry Systems* 40: 159–176.

Jones, C. G., J. H. Lawton, and M. Shachak. 1994. Organisms as ecosystem engineers. *Oikos* 69: 373–386.

Jones, C. G., J. H. Lawton, and M. Shachak. 1997. Positive and negative effects of organisms as physical ecosystem engineers. *Ecology* 78: 1946–1957.

Kikvidze, Z., and R. M. Callaway. 2009. Ecological facilitation may drive major evolutionary transitions. *BioScience* 59: 399–404.

Knops, J. M. H., D. Tilman, N. M. Haddad, S. Naeem, C. E. Mitchell, J. Haarstad, M. E. Ritchie, et al. 1999. Effects of plant species richness on invasion dynamics, disease outbreaks, insect abundances, and diversity. *Ecology Letters* 2: 286–293.

Laland, K. N., F. J. Odling-Smee, and M. W. Feldman. 1996. On the evolutionary consequences of niche construction. *Journal of Evolutionary Biology* 9: 293–316.

Laland, K. N., F. J. Odling-Smee, and M. Feldman. 1999. Evolutionary consequences of niche construction and their implications for ecology. *Proceedings of the National Academy of Sciences* 96: 10242–10247.

Lenz, T. I., and J. M. Facelli. 2003. Shade facilitates an invasive stem succulent in a chenopod shrubland in South Australia. *Austral Ecology* 28: 480–490.

Loreau, M., S. Naeem, P. Inchausti, J. Bengtsson, J. P. Grime, A. Hector, D. U. Hooper, et al. 2001. Biodiversity and ecosystem functioning: Current knowledge and future challenges. *Science* 294: 804–808.

Maron, J. L., and P. G. Connors. 1996. A native nitrogen-fixing shrub facilitates weed invasion. *Oecologia* 105: 302–312.

Maron, J. L., and R. L. Jefferies. 1999. Bush lupine mortality, altered resource availability, and alternative vegetation states. *Ecology* 80: 443–454.

Menchaca, L., and J. Connolly. 1990. Species interactions in white clover-ryegrass mixtures. *Journal of Ecology* 78: 223–232.

Miller, T. E. 1994. Direct and indirect species interactions in an early oldfield plant community. *American Naturalist* 143: 1007–1025.

Moore, P. D. 1990. Vegetation's place in history. *Nature* 347: 710.

Mulder, C. P. H., D. D. Uliassi, and D. F. Doak. 2001. Physical stress and diversity-productivity relationships: The role of positive interactions. *Proceedings of the National Academy of Sciences* 98: 6704–6708.

Naeem, S., K. Hakansson, J. H. Lawton, M. J. Crawley, and L. J. Thompson. 1996. Biodiversity and plant productivity in a model assemblage of plant species. *Oikos* 76: 259–265.

Nicolson, M., and R. P. McIntosh. 2002. H. A. Gleason and the individualistic hypothesis revisited. *Bulletin of the Ecological Society of America* 83: 133–142.

Odling-Smee, F. J., K. N. Laland, and M. W. Feldman. 1996. Niche construction. *American Naturalist* 147: 641–648.

Padilla, F. M., and F. I. Pugnaire. 2006. The role of nurse plants in the restoration of degraded environments. *Frontiers in Ecology and the Environment* 4: 196–202.

Palaniappan, V. M., R. H. Marrs, and A. D. Bradshaw. 1979. The effect of *Lupinus arboreus* on the nitrogen status of china clay wastes. *Journal of Applied Ecology* 16: 825–830.

Patten, D. T. 1978. Productivity and production efficiency of an upper Sonoran Desert ephemeral community. *American Journal of Botany* 65: 891–895.

Pickart, A. J., L. M. Miller, and T. E. Duebendorfer. 1998. Yellow bush lupine invasion in northern California coastal dunes, I: Ecological impacts and manual restoration techniques. *Restoration Ecology* 6: 59–68.

Pugnaire, F. I., and M. T. Luque. 2001. Changes in plant interactions along a gradient of environmental stress. *Oikos* 93: 42–49.

Ratliff, R. D., D. A. Duncan, and S. E. Westfall. 1991. California oak-woodland overstory species affect herbage understory: Management implications. *Journal of Rangeland Management* 44: 306–310.

Rumbaugh, M. D., D. A. Johnson, and G. A. Van Epps. 1982. Forage yield and quality in a Great Basin shrub, grass, and legume pasture experiment. *Journal of Range Management* 35: 604–609.

Schade, J. D., R. Sponseller, S. L. Collins, and A. Stiles. 2003. The influence of *Prosopis* canopies on understorey vegetation: Effects of landscape position. *Journal of Vegetation Science* 14: 743–750.

Siemann, E., and W. E. Rogers. 2003. Changes in light and nitrogen availability under pioneer trees may indirectly facilitate tree invasions of grasslands. *Journal of Ecology* 91: 923–931.

Simberloff, D., and B. Von Holle. 1999. Positive interactions of nonindigenous species: Invasional meltdown? *Biological Invasions* 1: 21–32.

Spehn, E. M., M. Scherer-Lorenzen, B. Schmid, A. Hector, M. C. Caldeira, P. G. Dimitrakopoulos, J. A. Finn, et al. 2002. The role of legumes as a component of biodiversity in a cross-European study of grassland biomass nitrogen. *Oikos* 98: 205–218.

Thomas, B. D., and W. D. Bowman. 1998. Influence of N_2-fixing *Trifolium* on plant species composition and biomass production in alpine tundra. *Oecologia* 115: 26–31.

Tilman, D., J. Knops, D. Wedin, P. Reich, M. Ritchie, and E. Siemann. 1997. The influence of functional diversity and composition on ecosystem processes. *Science* 277: 1300–1302.

Tilman, D., P. B. Reich, J. Knops, D. Wedin, T. Mielke, and C. Lehman. 2001. Diversity and productivity in a long-term grassland experiment. *Science* 294: 843–845.

Tilman, D., D. Wedin, and J. Knops. 1996. Productivity and sustainability influenced by biodiversity in grassland ecosystems. *Nature* 379: 718–720.

Titus, J. H., and S. Tsuyuzaki. 2003. Influence of a non-native invasive tree on primary succession at Mt. Koma, Hokkaido, Japan. *Plant Ecology* 169: 307–315.

Turkington, R. 1989. The growth, distribution and neighbour relationships of *Trifolium repens* in a permanent pasture, V: The coevolution of competitors. *Journal of Ecology* 77: 717–733.

Turkington, R., and L. A. Mehrhoff. 1990. The role of competition in structuring pasture communities. In *Perspectives on plant competition*, ed. J. B. Grace and D. Tilman, 308–366. New York: Academic Press.

Valiente-Banuet, A., and M. Verdú. 2008. Facilitation can increase the phylogenetic diversity of plant communities. *Ecology Letters* 10: 1029–1036.

Vitousek, P. M., and L. R. Walker. 1989. Biological invasion by *Myrica faya* in Hawaii: Plant demography, nitrogen fixation, and ecosystem effects. *Ecological Monographs* 59: 247–265.

Vitousek, P. M., L. R. Walker, L. D. Whiteaker, D. Mueller-Dombois, and P. A. Matson. 1987. Biological invasion by *Myrica faya* alters ecosystem development in Hawaii. *Science* 238: 802–804.

Vivanco, J. M., H. P. Bais, F. R. Stermitz, G. C. Thelen, and R. M. Callaway. 2004. Biogeographical variation in community response to root allelochemistry: Novel weapons and exotic invasion. *Ecology Letters* 7: 285–292.

Walker, L. R., and P. M. Vitousek. 1991. An invader alters germination and growth of a native dominant tree in Hawai'i. *Ecology* 72: 1449–1455.

Went, F. W. 1942. The dependence of certain annual plants on shrubs in southern California deserts. *Bulletin of the Torrey Botanical Club* 69: 101–114.

Zhang, F., and L. Li. 2003. Using competitive and facilitative interactions in intercropping systems enhances crop productivity and nutrient-use efficiency. *Plant and Soil* 248: 305–312.

2 Plant Interaction Indices Based on Experimental Plant Performance Data

Zaal Kikvidze and Cristina Armas

CONTENTS

2.1 INTRODUCTION

Plant interaction is a central concept for plant ecology, and the success of our research depends greatly on our ability to accurately quantify these interactions. However, this is not an easy task, as plant interactions entirely reflect the diversity, complexity, and variability of natural systems. Consequently, an array of quantitative tools reflecting many aspects of plant interactions has been developed, and efforts are continually made to improve these tools. In this chapter we will analyze and discuss current quantitative tools of widespread use in plant ecology, as well as propose and discuss some new ones.

Life is a process of interactions of an organism with its environment, including other living and nonliving objects. Any effect of an organism on another is an example of interaction. Consequently, ecological interactions may be of many different kinds and exhibit various aspects. Ecology textbooks usually devote significant

TABLE 2.1

Simple Classification of Ecological Interactions Based on Whether an Effect of One Organism on Another Is Positive (+), Neutral (0), or Negative (−)

		Effect on Species 2		
		+	0	−
	+	Mutualism Facilitation	Commensalism Facilitation	Predation Herbivory Parasitism
Effect on Species 1	0	Commensalism Facilitation	Neutralism	Competition
	−	Predation Herbivory Parasitism	Competition	Competition

Source: Based on van Andel (2005).

space to the introduction, classification, and discussion of interactions, and almost all use the elegant +, 0, − signs classification (Table 2.1, based on van Andel 2005). As we see, the first approach to interactions is to classify whether the effects of the interacting organisms are positive, negative, or neutral. Commensalism, facilitation, and mutualism are examples of positive interaction, while competition, predation, herbivory, and parasitism include negative interactions (Table 2.1). The absence of effects is usually understood as neutralism (the central cell of Table 2.1). Neutralism, however, does not necessarily mean an absence of interactions. Because we usually measure the net outcome of interactions among organisms, populations, or species, the observed neutral outcome may result from counterbalancing positive and negative effects.

Plants are sessile and consume similar resources, and, as a consequence, competition for resource preemption and facilitation through stress amelioration are by far the most common types of interactions found among plants. In this chapter we analyze the quantitative tools that were developed primarily for these two types of interaction. An overwhelming majority of studies measure plant interactions in terms of biomass accumulation rate in the presence versus absence of other plants, and far fewer studies analyze other aspects of plant population success, such as fecundity and survival (Aarssen and Keogh 2002), although they are important components of plant fitness. This chapter reflects the existing situation and, therefore, synthesizes primarily the works based on plant growth analysis. Likewise, we will not discuss quantitative measures for less-studied interactions such as plant parasitism and allelopathy.

2.2 EXPERIMENTAL APPROACHES FOR STUDYING PLANT INTERACTIONS

Historically, competition was thought to be the major and ecologically most important type of interaction (see also Chapter 1, this volume). To study the mechanisms and the role of competition, experimental approaches were developed both for controlled conditions in laboratories/greenhouses and outdoor/field plots. In these experiments, density of plant individuals, species composition, and abundance were manipulated, and favorable conditions were maintained by watering and even fertilizing. Additionally, plots were protected against grazers and pests using fencing and pesticides. When facilitation started to attract the attention of researchers about two decades ago, it was studied basically in field conditions.

The basic experimental approaches for studying competition are presented in Table 2.2. We roughly divide experiments into two groups: those with controlled conditions and field experiments (see above). These experiments can be focused on a single species, on interactions between two (pairs) or among a few species, or applied to multispecies systems. Next we will discuss briefly these types of experimental approaches as well as their advantages and disadvantages. The neighbor-removal technique will be analyzed in more detail as this approach can be applied to various numbers of species under variable environmental conditions.

Plant performance is an important notion for experimental research on plants, including interactions. Plant performance must ultimately be measured as fitness of the population, yet for convenience it is usually measured as a growth (a rate of plant biomass accumulation). Biomass can be measured in various ways. One approach is noninvasive measurements where the biomass can be assessed by cover, height,

TABLE 2.2
Experimental Designs for Competition

	One Species	Two or a Few Species	Multispecies
Controlled	Intraspecific competition by density-dependence	Pairwise presence/ absence	Pairwise with a phytometer
		Resource uptake	Resource uptake
Field	Neighbor removal	Neighbor removal	Neighbor removal
	A phytometer	Two or a few phytometers	Multiple phytometers

Note: The experimental designs are classified depending on how many target species will be studied and whether the experiment is performed under controlled environmental conditions or in the field. The specific types of designs (e.g., neighbor removals, the use of phytometers, and resource-uptake measurements) are discussed in the text.

volume (i.e., projective cover multiplied by height), number of shoots, and length of largest shoot or number of leaves. Other, more precise measurements are destructive: plants are harvested and dried to constant dry mass and then weighed. Detailed descriptions of these techniques can be found in textbooks on plant ecology and physiology (e.g., Sala and Austin 2000).

2.2.1 NEIGHBOR EFFECTS ON ONE SPECIES: EXPERIMENTS IN CONTROLLED CONDITIONS

Experiments analyzing the mechanisms of competition, particularly those quantifying intraspecific competition, usually focus on one target species (e.g., Firbank and Watkinson 1990). A study can be focused on a single species for various reasons. For example, it may be a rare species under threat of extinction, and the design of a proper conservation program will require the knowledge about whether transplanting of plants in certain protected habitats will be successful. Second, a single species may attract interest because it is a noxious invader, and researchers need to know its population dynamics. Third, a single species can be of economical interest, e.g., for timber, and managers need to determine the optimal density of seedlings. Finally, a species can be used as a phytometer. Experiments with phytometers use key species grown under controlled conditions to test the effects of specific variables on plant growth (Wheeler, Shaw, and Cook 1992). A phytometer can also be used to assess the intensity and direction of plant interactions (e.g., Reader et al. 1994), and such experiments are essential to analyze the importance of plant interactions (Brooker et al. 2005 [see the section on importance and intensity of interactions]).

Studies analyzing intraspecific competition basically look at the density dependence of interactions. Logically, the more organisms competing for a limited resource, the more intense will be the competition for it. These experiments and their theory were developed as early as the 1950s and 1960s (Kira, Ogawa, and Sakazaki 1953; Shinozaki and Kira 1956; Watkinson 1980). In such experiments, plants are usually sown at different densities and their performance is studied by analyzing growth as a function of density.

Density-dependence experiments may be carried out in different environments. Usually plants are sown at different densities in pots with artificial soils. Alternatively, researchers use outdoor plots with controlled conditions: Soils are homogenized (ploughed, sieved, and treated with pesticides), watered during the experiment to avoid accidental mortality, and weeded regularly to prevent the influence of other species. These plots are also fenced to exclude grazers. After completing the experiment, results can be presented as plant growth responses along the range of their densities (Figure 2.1a). Plants usually show a pattern of declining performance with increasing density of individuals. This pattern can be quantified with the following model (Firbank and Watkinson 1990):

$$w = w_{m}(1 + aN)^{-b} \qquad (2.1)$$

where w is mean per capita dry weight, w_{m} is maximum mean per capita dry weight at lowest density, a is an area required to achieve maximum dry weight (w_{m}), N is the density of plants (number of individuals per pot or per unit area), and b is the

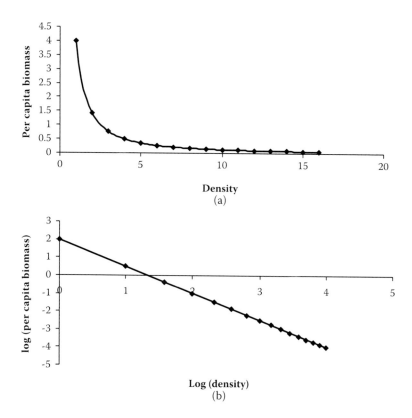

FIGURE 2.1 Intraspecific competition depicted as a density dependence of plant performance (measured as growth or per capita biomass accumulation rate).

efficiency of resource utilization. The parameters a and b describe how strongly the interactions depend on plant density.

In this model, the experimental population is even-aged, and at the early stages plants grow more or less uniformly. However, as the effects of competitive interactions among individuals become apparent, plants split into winners and losers. Such differentiation takes place sooner in denser populations. The loser individuals eventually die, a process known as self-thinning. Consequently, at later stages density-dependent mortality can characterize the process of competition. In other words, the values of a and b parameters change over time as plants grow. To quantify this process of self-thinning, Yoda et al. (1963) proposed the following equation:

$$w = cN^{-k} \tag{2.2}$$

where c is a constant factor that varies from species to species and k has a value of 3/2 for a wide range of species (White 1985). The curvilinear relationships described by Equation (2.2) (Figure 2.1a) can be transformed into a linear dependence using logarithms, for example, as:

$$\log_2 w = \log_2 c - k \log_2 N$$

Then, the known log-transformed values of N (independent variable) and w (dependent variable) can be regressed to calculate the parameter c (Figure 2.1b). This equation has many applications, e.g., in forestry for predicting the course of self-thinning in planted monospecific stands. For details, see White (1985).

2.2.2 Neighbor Effects on One Species: Field Experiments

In this approach, researchers seek to find the effects of neighbors on a target species, which may be a resident species of ecological or economical interest, or a phytometer. These experiments compare the performances of target individuals growing in the presence versus the absence of neighbors (Reader and Best 1989). The neighbor-removal experiment is a simplified, broad approach that does not differentiate between intra- and interspecific competition, nor does it make a distinction between what species are removed around target individuals. By using many replicates, the neighbor-removal approach gives a good average estimate of the effects that neighbor plants have on one another. These experiments may be considered an extreme case of density manipulations: no neighbors versus intact neighborhood. Such generality gives neighbor-removal experiments the advantage of being useful in many designs. As shown in Table 2.2, neighbor-removal experiments are commonly used with any number of species. These applications of neighbor-removal experiments with different numbers of species are described later in this chapter. Here we will give an example explaining a neighbor-removal experiment in a herbaceous community, where this approach is regularly used.

The interaction effects of neighbors on the target species are assessed by comparing target plant performance with that of paired control plants in which neighbors are left intact. For example, the experiment can start by marking 12 triplets of target individuals of the experimental species, selecting them to be as similar in size as possible (same shoot height, same number of leaves, etc.) at the beginning of the growing season. From each of these 12 triplets, one target is selected and neighbors are removed around it, while the second one is marked and left as a control. The third individual of the triplet is harvested for assessing the initial biomass of plant individuals (see Section 2.4), dried to constant dry mass (three days at 70–80°C), and weighed. Plants of each treated-control-harvested triplet are usually selected as close as possible, and within the same apparent microenvironment, but apart enough so that they are unlikely to influence one another (20–40 cm). At the end of the experiment, the aboveground parts of the control and experimental plants are harvested, dried to constant dry mass, and weighed. The analyses of the obtained biomass data are described in the following subsection.

2.2.3 Data Analyses and Interaction Indices

Data analyses follow primarily two consecutive steps: The first is to test whether there are statistically significant responses of plants to experimental manipulations. If the responses are proved to be significant, then the second step will be to quantify them with an appropriate tool.

The first step analyzes the mean and variance of, for example, biomass values to test whether growth was different between treated and control individuals. In the previously described example, treated and control individuals were paired; consequently, a paired t-test seems to be the most appropriate tool. If the experimental design is pooled instead of paired, a regular t-test can be applied. If the experiment deals with more than one species, then ANOVA (analysis of variance) is recommended. Overall, this first step is performed to find whether experimental manipulation has led to a measurable (statistically significant) effect using raw data and standard statistical tests. This issue is important because the quantitative approaches applied later imply data conversions that are often not ideal for standard tests (see Section 2.3).

If the effect of experimental manipulation is significant, we can proceed further and apply quantitative tools to measure the intensity of this effect. To quantify plant performance, we need to convert plant biomass data into the growth rate that occurred during the experimental period (i.e., from the start of the manipulative treatment to the end of the experiment). One problem may be that growth rate measurements can be expressed in two ways: absolute growth rate or relative growth rate.

Absolute growth rate (AGR) is defined as the increase in plant mass, M, over a period of time, t: $\text{AGR} = dM/dt$, or, in an integral form,

$$\text{AGR} = (M_{\text{Final}} - M_0)/t_{\text{Exp}} \tag{2.3}$$

where M and t are mass and time in general, M_{Final} and M_0 are final and initial biomasses of target plants, respectively, and t_{Exp} is the duration of experiment in time units (days, months, years, etc.). The absolute growth rate exactly measures how many mass units have been accumulated per unit of time. However, this equation has one caveat: the initial biomass M_0 can vary from small to huge, but it does not affect the absolute growth rate result. Intuitively, plants that grow from smaller initial size perform better than plants that grow with the same absolute rate but start from larger size.

This concern is reflected in the relative growth rate (RGR), which incorporates initial biomass into the differential equation: $\text{RGR} = dM/(Mdt)$. The integral of this equation will be $M_{\text{Final}} = M_0 \exp(\text{RGR}\ t_{\text{Exp}})$. After taking the natural logarithm of both sides of the last equation, a few simple algebraic conversions will produce:

$$\text{RGR} = (\ln M_{\text{Final}} - \ln M_0)/t_{\text{Exp}} \tag{2.4}$$

RGR is largely used in experimental studies on plant ecology (Poorter and Garnier 1999) as well as in some studies on plant interactions (e.g., Reader et al. 1994). The advantage of RGR is that it is a relative measure and thus is useful for comparing different species in different situations.

Researchers started to quantitatively study facilitative interactions later than competition. However, there were not only conceptual biases that held back the study of facilitation. Other considerable technical issues also caused positive interactions to escape our attention. The first studies on plant interactions were set under "standard" environmental conditions quite favorable for plant productivity and free from environmental stresses. In such conditions, competition usually predominates. At the same time, researchers were not able to reproduce in the laboratory the stressful conditions under which facilitation is likely to occur or predominate. Accordingly, the main experimental approach to analyze facilitation was field studies, very often based on neighbor-removal experiments. There were also certain biases among popular methods used for measuring interactions, which made it significantly more difficult to detect facilitation than competition. These biases are still undervalued (Kikvidze, Armas, and Pugnaire 2006; Connolly, Wayne, and Bazzaz 2001; we discuss this problem further in Section 2.4). Finally, facilitative and competitive mechanisms are not similar and may operate at different temporal scales (Callaway and Pugnaire 2007), making the efforts to detect facilitation together with competition less probable.

Facilitation can be defined theoretically as reversed competition: A target plant must perform better in the presence of other plants than it does alone. Indeed, in the case of competition, neighbor removal will lead to improved performance of treated individuals compared with controls with intact neighbors: $P_{n-} > P_{n+}$, where P_{n-} and P_{n+} are, respectively, performance in the absence and presence of neighbors, measured, for example, as AGR. Conversely, when facilitation occurs: $P_{n-} < P_{n+}$. One of the simplest ways to quantify and compare these situations is using the ratio:

$$P_{n+}/P_{n-} = \alpha \qquad (2.5)$$

If there is no response to neighbor removal, α will be equal to 1, whereas $\alpha < 1$ or $\alpha > 1$ when plant performance after neighbor removal, respectively, increases (competition) or decreases (facilitation). The ratio α is always positive and has no upper limit, as it may increase to infinity when facilitation is crucial for survival, i.e., the case of obligatory mutualism. These mathematical properties of α are not ideal for an index, and thus there were other indices developed to quantify plant interactions (see reviews in Goldberg et al. 1999; Weigelt and Jolliffe 2003). Many of these indices can be expressed with the help of α. For example, one of the most popular indices, lnRR (log response ratio; Hedges, Gurevitch, and Curtis 1999), is a simple log transformation of α:

$$\ln RR = -\ln \alpha$$

$\ln RR = 0$ when there are no effects detectable ($\alpha = 1$), and increases or decreases, respectively, with growing intensity of competition ($\alpha < 1$) or facilitation ($\alpha > 1$). This index was recommended for its favorable statistical properties, which are convenient for meta-analysis (Goldberg et al. 1999), but it does not have a maximum and minimum, cannot be calculated when there is competitive exclusion (i.e., when $P_{-n} = 0$), and is based on a geometric mean, which is unusual (Oksanen, Sammul, and Grünthal 2006).

Another index, RNE (relative neighbor effect), was suggested by Markham and Chanway (1996) and became popular, especially with neighbor-removal experiments. This index has different equations for competition and facilitation. Here we present it in the modified form after Callaway et al. (2002). For competition:

$$RNE = (P_{n+} - P_{n-})/P_{n-} = \alpha - 1 \qquad (2.6a)$$

For facilitation:

$$\text{RNE} = (P_{n+} - P_{n-})/P_{n+} = 1 - 1/\alpha \qquad (2.6b)$$

RNE values range from -1 ($\alpha < 1$, competition) through 0 ($\alpha = 1$, no interactions or no effects detectable) to $+1$ ($\alpha > 1$, facilitation). Despite its popularity, this index has several mathematical and statistical caveats (Goldberg et al. 1999; Armas, Ordiales, and Pugnaire 2004; Oksanen, Sammul, and Grünthal 2006). Among other problems, RNE requires logical conversions when interactions shift from positive to negative and vice versa. This requirement not only complicates the calculations, but also makes RNE an inconvenient and undesirable index for mathematical models (Armas, Ordiales, and Pugnaire 2004).

A recently suggested index is RII (relative interaction index; Armas, Ordiales, and Pugnaire 2004):

$$\text{RII} = (P_{n+} - P_{n-})/(P_{n+} + P_{n-}) = (\alpha - 1)/(\alpha + 1) \qquad (2.7)$$

RII has defined limit values, which range from -1 ($\alpha < 1$) through 0 ($\alpha = 1$, no interactions or no effects detectable) to $+1$ ($\alpha > 1$, facilitation). Armas et al. (2004) compared the statistical and mathematical properties of RII, RNE, and lnRR and concluded that RII is the most recommendable index of the intensity of plant–plant interactions. As we see, the above indices are mathematically closely related through α. Therefore, α can help compare the data expressed with different indices.

2.2.4   INTERACTIONS AMONG TWO OR A FEW SPECIES

In the previous section we described some quantitative tools applied to a single target species. This section briefly presents other approaches developed for experiments with two or more species.

The mechanisms of interspecific competition have been studied often in experiments based on pairwise tests of competition. These studies combine experiments in the laboratory and in the field. The pairwise approach is very restrictive on the number of species that can be simultaneously studied because the number of pairs to be tested increases with the quadratic power of the number of species to be analyzed. For this reason, only two or a few species usually can be tested simultaneously. Pairwise experiments are first conducted under standard laboratory conditions to establish a "competitive hierarchy," and then field experiments test whether this competitive hierarchy coincides with the abundance of ranked species (Keddy 1990). *Hierarchy* here means ranking the species by their competitive ability: from stronger to weaker competitors. Although competition indices used in those works did not become popular, some approaches continue to be important in plant interaction studies. These approaches are based on special experimental designs of density manipulations. Among them, the most popular are additive series, replacement series, response surface, and bivariate factorial designs (Figure 2.2). These approaches combine intra- and interspecific interactions; essentially, they are

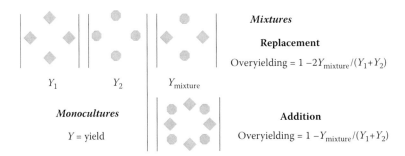

FIGURE 2.2 Designs of pairwise competition/overyielding experiments. Monocultures of two species are shown in the left upper corner (squares and circles). In mixture plots, plant density can be kept constant when individuals of one species are replaced by individuals of other species (upper panel, replacement series). Another way is increasing density twice, adding another species (lower panel, addition series). The equations show overyielding is calculated for these two approaches. If there are no interactions, the ratio of yields will equal 1 (i.e., no or zero overyielding).

experiments on *overyielding*, a measure that compares the mean yield obtained in monospecific stands to the yield obtained in mixed communities. Briefly, the aim is to quantify whether there is an increase in productivity with increasing richness, and it is an important tool in studies on the effects of biodiversity on productivity (Hooper et al. 2005). The equation selected to measure overyielding depends on the particular experimental design: whether it is a replacement or additive series design (Figure 2.2). Response–surface and bivariate factorial designs are modifications of these experimental designs, in which densities vary systematically or at random (Connolly, Wayne, and Bazzaz 2001; Reynolds and Rajaniemi 2007). A similar approach is the accurate observation of a population dynamics in nature and then using nonlinear regression methods to analyze intraspecific and interspecific competition. Rees, Grubb, and Kelly (1996) successfully used such an approach, although it requires very laborious recordings and is based on correlative and not experimental data.

As previously mentioned, the described designs do not allow the study of large sets of species. To overcome this problem, researchers use special approaches such as pairwise experiments with a common phytometer (the most abundant species thought to be the top competitor in the hierarchy). Another limitation of such experiments is that they are conducted under "standard environmental conditions." In fact, such conditions are rare in nature, and the studies restricted by these conditions fail to address the important question of how plant interactions depend on environmental conditions.

Neighbor-removal experiments have the advantage of being free from most of the limitations mentioned above, and this approach is often used to study competition among two to three species in grasslands (Reynolds and Rajaniemi 2007). These species usually are codominants belonging to different growth forms, e.g., grasses, legumes, and forbs.

2.2.5 MULTISPECIES EXPERIMENTS

As mentioned in the previous section, pairwise experiments are very restrictive for studying multispecies interactions, and several approaches have been suggested to reduce the number of necessary experiments. (See Reynolds and Rajaniemi 2007 for a review.) We will cite some of them very briefly.

The first approach is overyielding in a multispecies mixture. In this experimental design, the performance of each species grown in a single mixed multispecies community is compared with the performance of each species in monocultures, analogously to pairwise designs (additive and replacement series, Figure 2.2; Campbell and Grime 1992; Turkington, Klein, and Chanway 1993). The replacement technique is the one often used, in which overyielding is calculated as:

$$\text{Overyield} = 1 - nY_{\text{mixture}}/(Y_1 + Y_2 + \cdots + Y_i + \cdots + Y_n) \tag{2.8}$$

where Y_{mixture} is yield obtained in the mixed community, n is the number of species, and Y_i is yield obtained in monoculture of the ith species.

An alternative approach is to remove species from the existing communities (e.g., Gurevitch and Unnasch 1989; Keddy 1989a, 1989b) and compare the performance of the remaining species with that in the no-removal controls. There is a modification of this design in which densities of species change but their initial relative abundances are kept constant. Then, the final relative abundance is compared with the initial relative abundance of each species performing an analysis of residuals of the corresponding regression line (Goldberg, Turkington, and Olsvig-Whittaker 1995).

A rather different approach is used in resource competition studies (Tilman 1990). This approach assumes that the most competitive species in a community is the one that is able to reduce the limiting resource to the lowest level (Figure 2.3). In other words, this approach assumes that the species that reduces the limiting resource to the lowest level in monoculture will eventually be able to displace all competitors. In this case, the quantitative index is R*, defined as the level to which the individuals of a population deplete the limiting resource at equilibrium (Tilman 1990). R* can be

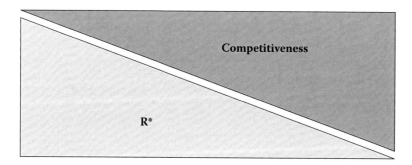

FIGURE 2.3 Schematic representation of the resource competition model. "Competitiveness" is measured as the ability of a species to deplete the limiting resource. R* is the level to which a species can deplete this resource. Consequently, R* is negatively proportional to the competitive ability of a plant.

determined experimentally, fitting data with a system of differential equations that describe growth and resource acquisition (Tilman 1990). Another assumption of this approach is that species abundance in a given community is determined by its competitive ability (as measured by R*). Therefore, the species-abundance ranking in a community can be predicted from the resulting R* values (Reynolds and Rajaniemi 2007) and then compared with the observed abundance ranking. The advantage of this approach is that it can be applied to multispecies systems, and that it is based on a certain mechanism of competition, i.e., the depletion of a resource to an extent that it becomes limiting for plant species.

The last approach covered here is neighbor-removal experiments, already discussed in preceding sections. This approach can be applied to any reasonable number of species of interest in a given community. The approach is the same as that used for two or few species, except that the number of target species will increase. The disadvantage of the neighbor-removal approach is that it does not distinguish intra- and interspecific competition or direct and indirect effects. But this disadvantage is more than compensated for by the fact that neighbor removal is by far the most convenient method for studying multiple species in the field and along environmental gradients. Moreover, this technique allows tracking changes in the direction, intensity, and even the importance of interactions, and therefore it has been especially helpful for detecting and quantifying positive interactions in the field.

2.3   IMPORTANCE AND INTENSITY OF PLANT INTERACTIONS

The previously discussed indices of interactions measure the intensity of competitive and facilitative effects of plants on one another, thereby quantifying the change in plant performance due to the effects of plant interactions. However, plant performance or fitness depends on many other factors other than the presence/absence of interacting neighbors. The physical environment as well as the strength and frequency of stress factors (drought, predation, disturbances, etc.) will all affect plant fitness. Therefore, the same intensity of plant interactions under benign versus severe environments may contribute differently to plant performance relative to other factors. This led to the definition of the importance of the interaction in contrast to the intensity of the interaction, which can be best expressed graphically (Figure 2.4, modified from Welden and Slauson 1986; Brooker et al. 2005). In case 1, the total reduction in performance of target species due to competition and other factors is 50 arbitrary units, of which competition accounts for 25 units. The intensity of competition is therefore 25 units, while the importance of competition (the impact of competition as a proportion of the total impact of the environment) is $25/(25 + 25) = 0.50$. In case 2, although the intensity of competition is the same, i.e., 25 units, the impact of other factors is now considerably greater (50 units). Therefore, the importance of competition is reduced to $25/(25 + 50) = 0.33$. Based on this idea, Brooker et al. (2005) defined the importance of competition as the impact of interactions relative to the impact of all the factors in the environment on plant performance, and derived an equation for quantifying competition importance:

$$\text{Competition importance} = (P_{n+} - P_{n-})/(P_{max} - P_{n+}) \qquad (2.9)$$

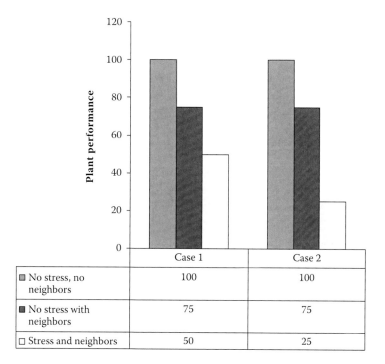

	Case 1	Case 2
▣ No stress, no neighbors	100	100
■ No stress with neighbors	75	75
☐ Stress and neighbors	50	25

FIGURE 2.4   Graphical illustration of the difference between the intensity and importance of competition (based on Welden and Slauson 1986). The figure shows the physiological status of one species growing in two sites (case 1 and 2) and under three different hypothetical conditions: optimum growth (light gray), when plant physiological status is only affected by competition with neighbors (dark gray), and when plant status is affected by the combined effect of competition and other environmental factors. In case 2, the effect of environmental stress on plant physiological status is two times more than in case 1.

where P_{max} is the maximum value of P_{n-} along the gradient, i.e., the performance of the plants growing without neighbors in the most productive, low-stress point of the gradient. Therefore, this index expresses the impact of competition as a proportion of the impact of the total environment. The relation of Equation (2.9) to the index of competition intensity RNE is evident (see Equation [2.6a]; see also Brooker et al. 2005). Yet, as we have already noted, RNE is not the best index for measuring the intensity of interactions (see Section 2.2.3; see also Armas, Ordiales, and Pugnaire 2004). Besides, the use of Equation (2.9) for measuring the importance of facilitation poses certain problems. Brooker et al. (2005) suggest the following general equation as an index of interaction importance:

$$\text{Interaction importance} = (P_{n+} - P_{n-})/[P_{max} - \min(P_{n+}, P_{n-})]$$

which, in the case of facilitation, means:

$$\text{Facilitation importance} = (P_{n+} - P_{n-})/(P_{max} - P_{n-}) \tag{2.10}$$

However, Equation (2.10) contradicts the definition of importance, which requires one to divide the difference in performance between plants growing with and without neighbors ($P_{n+} - P_{n-}$) by the difference between (a) maximum performance without neighbors and without stress (P_{max}) and (b) performance under a full set of factors (with neighbors and stress; P_{n+}).

This contradiction can be solved with the approach that was used to derive RII (Armas, Ordiales, and Pugnaire 2004). Mathematically, let us redefine the impact of all factors in the environment on plant performance as a sum of two differences: $(P_{max} - P_{n+}) + (P_{max} - P_{n-})$. The first term is the difference between the maximum performance of the isolated plant of the target species on the gradient (without neighbors) and its performance under a full set of factors (with neighbors and stress); the second term is the difference between the maximum performance of the isolated plant of the target species on the gradient and its performance under a given level of stress, but growing without neighbors. In other words, now we are assessing the importance of interactions as the impact of interactions relative to a *weighted* impact of all factors in the environment on plant performance. With this redefinition, the importance of interactions can be quantified as:

$$\text{RNI} = (P_{n+} - P_{n-})/(2P_{max} - P_{n+} - P_{n-}) \qquad (2.11)$$

where RNI stands for relative neighbor importance. The RNI retains all the key features of the importance index derived by Brooker et al. (2005).

There are two main areas of debate surrounding the popular indices of plant interactions:

1. Most indices are ratios, and there has been general criticism of the use of ratios in ecological research. All ratios are limited in their suitability for standard statistical analyses (Jasienski and Bazzaz 1999). However, if sample size is big enough (i.e., $n > 5$), ratios can be rigorously tested by means of randomization tests (Manly 1997; Fortín and Jacquez 2000). In addition, randomization tests have proved to be effective in analyzing other nonratio indices that are difficult to test by standard statistical methods, such as spatial distribution indices (Roxburgh and Matsuki 1999) and abundance distributions (Wilson and Roxburgh 2001; Kikvidze and Ohsawa 2002).

2. Freckleton and Watkinson (1997a, 1997b, 1999, 2000) discuss other potential problems specific to interaction indices. They state that such indices are inherently flawed because they do not allow the partitioning of interspecific and intraspecific components of any competitive or facilitative impact of neighbor plants on the target individual. Therefore, when competition or facilitation is found to vary along a gradient, it is impossible to tell whether this change is due to an absolute change in the number of interactions experienced, or to a change in the relative equivalence of neighboring species. However, it has been counterargued that, despite these problems, the ratios still provide a reasonable measure of net interactive effects and that the alternative response surface analysis is not practicable in field experiments (Markham 1997; Peltzer 1999).

Although these debates remain formally unresolved, they have apparently had little impact on ongoing research, as researchers continue to refer to a large body of literature that has already been developed using ratios as quantitative tools for measuring plant interactions.

2.4   THE PROBLEM WITH INITIAL BIOMASS

Plant biomass is the most commonly used measure of plant performance. Most often, researchers try to clarify the effect of interactions by comparing the biomass of naturally growing plants (or transplants or planted phytometers) under control and manipulated conditions. Experiments usually start with individuals that are young enough to avoid previous environmental effects, but mature enough to ensure survival and correct identification if several species are used. As a consequence, target individuals in both treatment and control have a similar initial mass prior to experiments. At the end of the experiment, the plants are harvested and the performance of target individuals is assessed by their final dry mass. However, the initial biomass is often ignored in these analyses (Connolly, Wayne, and Bazzaz 2001), and such omission may conceal the treatment effects. Connolly was the first to address this problem (Connolly and Wayne 1996), yet researchers did not pay much attention to it. Kikvidze, Armas, and Pugnaire (2006) recently revisited this problem. Based on their work, we discuss how ignoring initial biomass can cause systematic bias in plant interaction analyses, particularly when neighbor-removal designs are used.

We use the same definitions for growth or rate of biomass accumulation as were used for the interaction indices. "Performance" and "growth" are used here as synonyms. For plants without neighbors, biomass accumulation or net increase in mass per unit time during the experiment can be described as:

$$G_{n-} = (M_{n-} - M_0)/t \qquad (2.12)$$

where G_{n-} is growth of individuals without neighbors; M_{n-} and M_0 are, respectively, their final and initial mass; and t is time. Likewise, the equation for individuals surrounded by neighbors will be:

$$G_{n+} = (M_{n+} - M_0)/t \qquad (2.13)$$

where G_{n+} is growth of individuals with neighbors and M_{n+} is their final biomass. M_0, their initial mass, should be similar to the initial biomass of individuals growing without neighbors. Equations (2.12) and (2.13) can be used to derive an index that measures neighbor effects (analogous to Equation [2.5]):

$$\alpha = G_{n+}/G_{n-} = (M_{n+} - M_0)/(M_{n-} - M_0) \qquad (2.14)$$

To calculate the true value of α we need to include the values of initial biomass into Equation (2.14). However, initial biomass is often ignored and calculations are based on final mass, which will produce an apparent value of α':

$$\alpha' = M_{n+}/M_{n-} = (\alpha G_{n-} \cdot t + M_0)/(G_{n-} \cdot t + M_0) = (\alpha + \beta)/(1 + \beta) \qquad (2.15)$$

where β is introduced to simplify the algebraic expression. $\beta = M_0/G_{n-} \cdot t = M_0/(M_{n-} - M_0)$ and, as a ratio, compares the initial biomass to the increment in biomass. β approaches 0 when $M_0 \ll (M_{n-} - M_0)$, that is, when there is a rapid plant growth during the experiment and the accumulated biomass is much larger than its initial value. In this case the bias will not affect measurements too much, as α' will be very close to α. However, in any other cases, the apparent value α' will depend on its true value α as well as on the value of the β ratio.

Using Equations (2.14) and (2.15), it is possible to calculate the deviation of the apparent value α' from its true value. Figure 2.5a shows the surface of the percentage of deviation of apparent value α' from its true value, where α ranges from 0.1 to 10 and β ranges from 0.05 to 1. An asymmetric bias in measuring plant interactions is evident. For example, consider $\beta = 1$ (when the mass increment $M_{n-} - M_0$ is comparable with the initial mass M_0 due to slow growth). Then, at $\alpha = 0.1$, corresponding to 10 times reduced growth of plants with neighbors compared with those without neighbors, the apparent value of α' becomes 50% lower than its true value (Figure 2.5a). However, at $\alpha = 10$, corresponding to 10 times increased growth of plants with neighbors compared with those without neighbors, the bias increases dramatically to 450%! The point where apparent and true values are equal occurs only when $\alpha = 1$ (net neutral or no neighbor effects, Equations [2.14] and [2.15]), but, if initial biomass is used in calculations (as in Equation [2.14]), this bias disappears completely.

2.4.1 IMPLICATIONS FOR INTENSITY INDICES

The above analysis shows that ignoring initial biomass causes a dramatic directional bias when plants without neighbors grow more slowly than those with neighbors (when $\alpha > 1$, i.e., facilitation predominates). Indeed, if plant interactions are facilitative, as often happens in harsh and unproductive environments, removing neighbors will reduce growth in these target plants compared with plants with neighbors. Again, we will have $\alpha > 1$ at higher values of β, in the most biased conditions (Figure 2.5a). This bias will corrupt the real trends of plant interactions along environmental gradients: When initial biomass is ignored, strong facilitative effects will be considerably harder to detect than similarly strong competitive interactions. We will demonstrate this directional bias using the example of RII. For the true value of RII we will have Equation (2.7):

$$RII = (\alpha - 1)/(\alpha + 1)$$

The apparent values, correspondingly, will be (after Equations [2.14] and [2.15]):

$$RII' = [(\alpha + \beta)/(1 + \beta) - 1] / [(\alpha + \beta)/(1 + \beta) + 1]$$

Using these equations, we can calculate the deviation of the apparent value of the index from its true value, as we did with α. Figure 2.5b shows the obtained surface of the percentage deviation of the apparent value of RII from its true value in the same range as with α (corresponding to the range of -0.82 to 0.82 of true RII). Again,

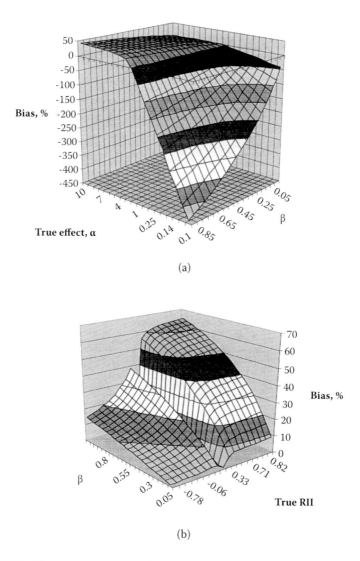

(a)

(b)

FIGURE 2.5 Bias in measurements of α (a) and RII (b) depends on their true value, and on growth response as measured by β (= ratio of initial and accumulated biomasses). Bias is calculated as the percentage deviation of apparent values from true values of α and RII.

the asymmetric bias is evident: At $\beta = 1$ and RII $= -0.82$ (or, when $\alpha = 0.1$, corresponding to 10 times increased growth of plants without neighbors compared with those with neighbors), the apparent value of RII becomes 8.12% lower than its true value. However, at RII $= 0.82$ (corresponding to $\alpha = 10$, or 10 times reduced growth of plants without neighbors compared with those with them), the bias increases to 64.5%. The point where apparent and true values are equal occurs at RII $= 0$ (neutral or no neighbor effects: $\alpha = 1$, Equations [2.14] and [2.15]). As we see, RII smooths the bias for competition, but for facilitation it is still unacceptably high.

We also analyzed other indices of competition and facilitation such as lnRR (Cahill 1999; Hedges, Gurevitch, and Curtis 1999) and RNE (Markham and Chanway 1996), which are mathematically related to RII (see Section 2.2.3). Not surprisingly, the analyses revealed a very similar bias related to the initial biomass. For facilitation, differences between true and apparent values reached values as high as 45% and 74% for RNE and lnRR, respectively (see also Kikvidze, Armas, and Pugnaire 2006). The lnRR index was often recommended due to its convenience for meta-analysis (Gurevitch et al. 1992; Hedges, Gurevitch, and Curtis 1999). However, when it is applied to unproductive conditions using only final biomass, the results of such analyses will be questionable, and we suggest reanalyzing this data set taking into account plant initial biomass.

In conclusion, initial biomass should be determined and used to measure plant performance in plant interaction experiments. However, in some cases, determination of initial mass may be impossible, for example in reanalyses of old data. In such cases, the use of some mathematical approaches can remove directional bias in a range of plant–plant interactions (see also Armas, Ordiales, and Pugnaire 2004 for details). However, even if directional bias is avoided, ignoring initial biomass can still reduce the probability of discovering significant effects, and this excessive conservatism should be taken into consideration.

2.4.2 RELATIONSHIP OF INTERACTION IMPORTANCE INDEX TO INITIAL BIOMASS

In Section 2.4, we showed that omission of the initial biomass in the quantification of the intensity of plant interactions may give biased results. But is this problem related to interaction importance? Looking at the index of the importance of interactions RNI, we will find that its mathematical form makes this index insensitive to the initial biomass problem. Actually, we can calculate RNI as:

$$\text{RNI} = (G_{n+} - G_{n-})/(2G_{\text{Max}} - G_{n+} - G_{n-})$$

(Growth values G here are synonymous with performance values P used in Equation [2.11]). Then, using Equations (2.14) and (2.15), we can write:

$$\text{RNI} = (M_{n+} - M_0 - M_{n-} + M_0)/(2M_{\text{Max}} - 2M_0 - M_{n+} + M_0 - M_{n-} + M_0)$$

$$= (M_{n+} - M_{n-})/(2M_{\text{Max}} - M_{n+} - M_{n-})$$

Consequently, RNI is unrelated to initial biomass, and we are free to ignore it or not. Clearly, as long as the determination of initial biomass is necessary for quantifying correctly the interaction intensity, it is simpler to use the same data for calculating the importance too—that is, using initial biomass in quantification of both aspects of interactions.

Here, we have used absolute growth rate (AGR, Equation [2.3]) for our analyses of the initial biomass problem. Performance measured as relative growth rate (RGR, Equation [2.4]) might be used with the same success, yet sometimes RGR produces negative values when the growth is too small. Negative values cannot be used to calculate RII or RNI properly. In such cases, the simple difference between the initial and final biomasses can be used for RII. For RNI, just final biomass values will be sufficient.

2.5 SUMMARY

Interaction among living organisms is a central concept of ecology, and the success of our research depends greatly on our ability to accurately quantify these interactions. This is not an easy task, as ecological interactions reflect the diversity, complexity, and variability of natural systems. Two basic purposes of conducting quantitative studies on plant interactions are: (a) understanding the mechanisms underlying the effects of interacting plants on each other, and (b) understanding the role of plant interactions in the processes that shape plant communities. This chapter described and discussed (a) different experimental approaches usually applied to study plant interactions and (b) some popular and current quantitative tools for analyzing the direction, intensity, and importance of plant interactions. Some of these approaches have been successfully used to study various aspects of plant interactions and establish important links between productivity, biodiversity, and spatial distributions from micro- to macroscales.

REFERENCES

Aarssen, L. W., and T. Keogh. 2002. Conundrums of competitive ability in plants: What to measure? *Oikos* 96: 531–542.

Armas, C., R. Ordiales, and F. I. Pugnaire. 2004. Measuring plant interactions: A new comparative index. *Ecology* 85: 2682–2686.

Brooker, R., Z. Kikvidze, F. I. Pugnaire, R. M. Callaway, P. Choler, C. J. Lortie, and R. Michalet. 2005. Importance of importance. *Oikos* 109: 63–70.

Cahill, J. F. 1999. Fertilization effects on interactions between above- and belowground competition in an old field. *Ecology* 80: 466–480.

Callaway, R. M., R. W. Brooker, P. Choler, Z. Kikvidze, C. J. Lortie, R. Michalet, L. Paolini, et al. 2002. *Nature* 417: 844–848.

Callaway, R. M., and F. I. Pugnaire. 2007. Facilitation in plant communities. In *Handbook of functional plant ecology*, 2nd ed., ed. F. I. Pugnaire and F. Valladares, 435–456. New York: Marcel Dekker.

Campbell, B. D., and J. P. Grime. 1992. An experimental testing of plant strategy theory. *Ecology* 73: 15–29.

Connolly, J., and P. Wayne. 1996. Asymmetric competition between plant species. *Oecologia* 108: 311–320.

Connolly, J., P. Wayne, and F. A. Bazzaz. 2001. Interspecific competition in plants: How well do current methods answer fundamental questions? *American Naturalist* 157: 107–125.

Firbank, L.G., and A. R. Watkinson. 1990. On the effects of competition: From monocultures to mixtures. In *Perspectives on plant competition*, ed. J. Grace and D. Tilman, 165–192. New York: Academic Press.

Fortín, M.-J., and G. M. Jacquez. 2000. Randomisation tests and spatially autocorrelated data. *Bull. Ecol. Soc. Am.* 81: 201–205.

Freckleton, R. P., and A. R. Watkinson. 1997a. Measuring plant neighbour effects. *Functional Ecology* 11: 532–534.

Freckleton, R. P., and A. R. Watkinson. 1997b. Response to Markham. *Functional Ecology* 11: 536.

Freckleton, R. P., and A. R. Watkinson. 1999. The mis-measurement of plant competition. *Functional Ecology* 13: 285–287.

Freckleton, R. P., and A. R. Watkinson. 2000. On detecting and measuring competition in spatially structured plant communities. *Ecology Letters* 3: 423–432.

Goldberg, D. E., T. Rajaniemi, J. Gurevitch, and A. Stewart-Oaten. 1999. Empirical approaches to quantifying interaction intensity: Competition and facilitation along productivity gradients. *Ecology* 80: 1118–1131.

Goldberg, D. E., R. Turkington, and L. Olsvig-Whittaker. 1995. Quantifying the community-level effects of competition. *Folia Geobotanica et Phytotaxonomica* 30: 231–242.

Gurevitch, J., L. L. Morrow, A. Wallace, and J. S. Walsh. 1992. A meta-analysis of competition in field experiments. *American Naturalist* 140: 539–572.

Gurevitch, J., and R. S. Unnasch. 1989. Experimental removal of a dominant species at two levels of soil fertility. *Canadian Journal of Botany* 67: 3470–3477.

Hedges, L. V., J. Gurevitch, and P. S. Curtis. 1999. The meta-analysis of response ratios in experimental ecology. *Ecology* 80: 1150–1156.

Hooper, D. U., F. S. Chapin, J. J. Ewel, A. Hector, P. Inchausti, S. Lavorel, J. H. Lawton, et al. 2005. Effects of biodiversity on ecosystem functioning: A consensus of current knowledge. *Ecological Monographs* 75: 3–35.

Jasienski, M., and F. A. Bazzaz. 1999. The fallacy of ratios and the testability of models in biology. *Oikos* 84: 321–326.

Keddy, P. A. 1989a. Effects of competition from shrubs on herbaceous wetland plants: A 4-year field experiment. *Canadian Journal of Botany* 67: 708–716.

Keddy, P. A. 1989b. *Competition*. London: Chapman and Hall.

Keddy, P. A. 1990. Competitive hierarchies and centrifugal organization in plant communities. In *Perspectives on plant competition*, ed. J. B. Grace and D. Tilman, 265–290. San Diego: Academic Press.

Kikvidze, Z., C. Armas, and F. I. Pugnaire. 2006. The effect of initial biomass in manipulative experiments on plants. *Functional Ecology* 20: 1–3.

Kikvidze, Z., and M. Ohsawa. 2002. Measuring the number of co-dominants in ecological communities. *Ecological Research* 17: 519–525.

Kira, T., H. Ogawa, and N. Sakazaki. 1953. Intraspecific competition among higher plants, 1: Competition-yield-density interrelationships in regularly dispersed populations. *Journal of the Institute of Polytechnics, Osaka City University* Series D4: 1–6.

Manly, B. F. J. 1997. *Randomisation, bootstrap and Monte Carlo methods in biology*. London: Chapman and Hall.

Markham, J. H. 1997. Measuring and modelling plant neighbour effects: Reply. *Functional Ecology* 11: 534–535.

Markham, J. H., and C. P. Chanway. 1996. Measuring plant neighbor effects. *Functional Ecology* 10: 548–549.

Oksanen, L., M. Sammul, and M. Grünthal. 2006. On the indices of plant-plant competition and their pitfalls. *Oikos* 112: 149–155.

Peltzer, D. A. 1999. Measuring plant neighbour effects in different systems. *Functional Ecology* 13: 283–284.

Poorter, H., and E. Garnier. 1999. Ecological significance of inherent variation in relative growth rate and its components. In *Handbook of functional plant ecology*, ed. F. I. Pugnaire and F. Valladares, 81–120. New York: Marcel Dekker.

Reader, R. J., and B. J. Best. 1989. Variation in competition along an environmental gradient: *Hieracium floribundum* in an abandoned pasture. *Journal of Ecology* 77: 673–684.

Reader, R. J., S. D. Wilson, J. W. Belcher, I. Wisheu, P. A. Keddy, D. Tilman, E. C. Morris, et al. 1994. Intensity of plant competition in relation to neighbor biomass: An intercontinental study with *Poa pratensis*. *Ecology* 75: 1753–1760.

Rees, M., P. J. Grubb, and D. Kelly. 1996. Quantifying the impact of competition and spatial heterogeneity on the structure and dynamics of a four-species guild of winter annuals. *American Naturalist* 147: 1–32.

Reynolds, H. L., and T. K. Rajaniemi. 2007. Plant interactions: Competition. In *Handbook of functional plant ecology*, 2nd ed., ed. F. I. Pugnaire and F. Valladares, 157–480. New York: Marcel Dekker.

Roxburgh, S. H., and M. Matsuki. 1999. The statistical validation of null models used in spatial association analyses. *Oikos* 85: 68–78.

Sala, O. E., and A. T. Austin. 2000. Methods of estimating aboveground net primary productivity. In *Methods in ecosystem science*, ed. O. E. Sala, R. B. Jackson, H. A. Mooney, and R. W. Howarth, 31–43. New York: Springer.

Shinozaki, K., and T. Kira. 1956. Intraspecific competition among plants, 7: Logistic theory of the C-D effect. *Journal of the Institute of Polytechnics, Osaka City University* Series D7: 35–72.

Tilman, D. 1990. Mechanisms of plant competition for nutrients: The elements of a predictive theory of competition. In *Perspectives on plant competition*, ed. J. Grace and D. Tilman, 117–141. New York: Academic Press.

Turkington, R., E. Klein, and C. P. Chanway. 1993. Interactive effects of nutrients and disturbance: An experimental test on plant strategy theory. *Ecology* 74: 863–878.

van Andel, J. 2005. Species interactions structuring plant communities. In *Vegetation ecology*, ed. E. van der Maarel. Oxford, UK: Blackwell Publishing.

Watkinson, A. R. 1980. Density-dependence in single-species populations of plants. *Journal of Theoretical Biology* 83: 345–357.

Weigelt, A., and P. Jolliffe. 2003. Indices of plant competition. *Journal of Ecology* 91: 707–720.

Welden, C. W., and W. L. Slauson. 1986. The intensity of competition versus its importance: An overlooked distinction and some implications. *Quarterly Review of Biology* 61: 23–44.

Wheeler, B. D., S. C. Shaw, and R. E. D. Cook. 1992. Phytometric assessment of the fertility of undrained rich-fen soils. *Journal of Applied Ecology* 29: 466–475.

White, J., 1985. *The self-thinning rule and its application to mixtures of plant population studies of demography*. London: Academic Press.

Wilson, J. B., and S. H. Roxburgh. 2001. Intrinsic guild structure: Determination from competition experiments. *Oikos* 92: 189–192.

Yoda, K., T. Kira, H. Ogawa, and K. Hozumi. 1963. Self-thinning in overcrowded pure stands under cultivated and natural conditions. *Journal of the Institute of Polytechnics, Osaka City University* Series D14: 107–129.

3 Consequences of Facilitation on Species Diversity in Terrestrial Plant Communities

Lohengrin A. Cavieres and Ernesto I. Badano

CONTENTS

3.1 INTRODUCTION

Determination of the processes that maintain biological diversity is one of the primary aims of community ecology (Morin 1999). Although early research focused on negative interactions (e.g., competition and predation) as the main biotic factors structuring plant communities and regulating biological diversity (e.g., Grime 1973; Connell 1978; Tilman 1982), more recent studies have started to examine the importance of positive interactions on community structure and biological diversity (Bruno, Stachowicz, and Bertness 2003; Michalet et al. 2006). Positive interactions are defined as nontrophic interspecific interactions that increase the average individual fitness of one species (Bertness and Callaway 1994; Callaway 1995; Bruno, Stachowicz, and Bertness 2003; Callaway 2007). Therefore, the presence of one plant species enhances the chances that another species co-occurs in the same place, indicating that positive interactions may determine biological diversity.

One obvious way in which positive interactions can increase species diversity is via the creation of habitat patches that are otherwise absent or rare due to the physical presence of a species (Jones, Lawton, and Shachak 1997). Positive interactions can also increase species diversity when the presence of one species modifies the environment in ways that reduce the frequency of some physical disturbance or physical stress, allowing less tolerant species to perform better (Hacker and Gaines 1997). The latter is an example of one of the most commonly recognized plant–plant positive interactions: facilitation by a nurse species, in which one species (the facilitator or nurse plant) provides shelter from physical stress or herbivory to other plant species (Callaway 2007). It has been suggested that positive interactions may have strong impacts in harsh environments, where the mitigation of extreme conditions by facilitator species can benefit many other species (Bertness and Callaway 1994; Callaway and Walker 1997; Brooker and Callaghan 1998).

Although some examples of nurse species have been reported in highly productive environments, such as tropical and subtropical forests (Powers, Haggar, and Fisher 1997), most of the examples are from harsh, low-productive habitats, such as deserts, cold alpine and arctic tundras, salt marshes, etc. (Callaway 1995, 2007). Harsh environments may restrict plants from acquiring resources, and any amelioration of these conditions will favor growth to the extent that it outweighs the negative, competitive impact of growing in close associations. Depending on the physical stress limiting the survival in a particular habitat, the mechanisms involved in the facilitation by a nurse plant species differ between habitats. For example, provision of shade, nutrients, and soil moisture by the nurse species have been the most common reported mechanisms involved in the facilitation in arid systems, while mitigation of strong wind and freezing temperatures have been proposed as the main mechanisms involved in facilitative interactions in arctic and alpine habitats. (For a complete review of mechanisms, see Callaway [2007] and Callaway and Pugnaire [1999].)

Shelter provided by nurse plants has been shown to increase the physiological performance, population density, and reproduction of some particular beneficiary species (Pugnaire et al. 1996a; Choler, Michalet, and Callaway 2001; Pugnaire and Luque 2001; Callaway et al. 2002; Maestre, Bautista, and Cortina 2003). However, despite these well-known effects of nurses at individual and population levels, the community-level consequences of these processes have received considerably less attention (Brooker et al. 2008).

Communities can be defined and described in very different ways (Morin 1999). Nonetheless, there are some community attributes that have been used extensively in the literature and that might help in the comparison among communities and within a given community at different stages. Some of the most used community attributes are species richness, diversity, species abundance relations, and species composition (Morin 1999).

The shelter provided by nurse plants may allow the persistence of some species that otherwise would be excluded from communities. Hacker and Gaines (1997) proposed a graphical model in which a facilitator species (nurse species) increases species richness under conditions of high mortality by ameliorating physical stress or by providing protection from disturbance or predation. Thus, the presence of nurses in a community would lead to a series of patterns in community attributes, including:

(a) higher species richness beneath nurses compared with areas outside them, and/ or (b) different composition of the plant assemblages growing beneath nurses and outside them. Nonetheless, changes in the microclimatic conditions beneath nurses may not produce changes in the overall composition of species growing beneath and outside nurses, but could generate changes in the abundance of some particular species, generating changes in species evenness. Nonetheless, nurses and their alternative habitats (open areas away from nurses) are part of the same community; hence the fact that nurse plants allow the persistence of some species that otherwise would be excluded from communities indicates that nurses are generating a net increase of the species diversity at the local scale when species assemblages growing within and outside nurses are integrated. Finally, the species sheltered by nurses could belong to different genera and/or families compared with those growing outside nurses. In this last case, nurses will be adding more than species richness at the community level: They will affect diversity at higher taxonomic levels, adding new lineages such as genera and families.

These community-level consequences of positive interactions are the focus of this chapter, and to illustrate them we will review the literature. Given that these types of questions seldom have been addressed, we will use our data set collected during the evaluation of the importance of a particular group of plants as nurses in the alpine environments of the Andes of southern South America to further illustrate some of the large- and small-scale consequences of facilitation in plant communities. In particular, we will focus on a series of simple questions such as: Are patches beneath nurses more diverse than outside them? Does species composition differ between these habitats? Do species assemblages show differential responses to the area of nurse and surrounding habitat patches? Despite the simplicity of these questions, a very low percentage of the studies focused on positive interactions among plants have addressed these issues. However, we find some examples that could help clarify the importance of facilitation for plant community diversity.

3.2 COMMUNITY-LEVEL EFFECTS OF NURSE SPECIES

3.2.1 SPECIES RICHNESS

The majority of the studies addressing some community-level consequences of facilitation have compared the diversity and species composition of the plant assemblages growing beneath nurses vs. plant assemblages growing outside nurses. This kind of study has been conducted in a wide range of ecosystems, although the majority have been carried out in stressful habitats. Nonetheless, Powers, Haggar, and Fisher (1997) conducted surveys of species diversity in plantations of seven different tree species in tropical forests in Costa Rica and observed that the number of woody species beneath canopies of five of these trees was 1.3–2.0 times higher than in control plots from deforested areas. Based on these results, they suggested that trees from plantations provide suitable habitat for shade-tolerant, late-successional woody species, hence facilitating the recovery of tropical rain forest in deforested sites.

Studies in arid systems are numerous, but with controversial results. For example, Raffaele and Veblen (1998) performed an experiment in the northern Argentinean

Patagonia, where they cut part of the canopy of the shrubs *Berberis buxifolia* and *Schinus patagonicus* and evaluated the species richness developed in cut vs. uncut sites beneath these nurses. They found that uncut sites contained higher species richness than cut areas, indicating that the provision of microhabitat with shade and soil moisture developed in uncut sites were key in these facilitative effects and their consequences on species richness. Larrea-Alcázar, López, and Barrientos (2005) reported that in the dry, high plains of the Bolivian Andes, the nurse tree species *Prosopis flexuosa* harbors higher species richness beneath its canopy compared with open areas away from such trees. Similarly, studies conducted in the Chilean *matorral*, which is characterized by a Mediterranean-type climate, have reported that pioneer shrub species (e.g., *Baccharis* spp.) act as nurses for herbaceous and late successional woody species, harboring higher species richness in areas within their reach than in areas where these nurse species are absent (Keeley and Johnson 1977; Fuentes et al. 1984; Badano et al. 2006).

Pugnaire, Armas, and Valladares (2004) conducted an extensive sampling in a semi-arid zone of southeastern Spain (Almería), where they compared the annual plant community that developed beneath six different shrub species with that developed outside them. Overall, they found that species richness beneath the different shrubs was similar to or lower than that found in gap areas between shrubs. Analyses of the standing biomass showed that some shrub species (e.g., *Thymus hyemalis*) inhibit the productivity of annual species beneath their canopy, whereas other shrubs (e.g., *Retama sphaerocarpa*) increased several times the productivity of annuals compared with open areas between shrubs. Interestingly, those shrub species that increased the biomass production contained lower species richness than the open areas away from them, indicating that facilitative effects do not necessarily translate into higher species richness in the microhabitat beneath nurses. Similarly, Rossi and Villagra (2003) studied the effects of *Prosopis flexuosa* in the arid zone of the Argentine Monte and reported that there were no significant differences in the number of species growing beneath nurses and in open areas away from them (see also Gutiérrez et al. 1993; de Villiers, Van Rooyen, and Theron 2001). Therefore, no clear trends of the effects of facilitative interactions on species richness at the patch scale can be discerned for arid and semi-arid habitats, which could be related to the different mechanisms, plant traits, and species-specific responses to the presence of neighbors found in these habitats (Michalet 2006, 2007).

Evidence from studies performed in alpine habitats also shows contrasting results. While some studies report higher species richness within nurses compared to open areas (e.g., Badano et al. 2002; Cavieres et al. 2002), others found either no differences (e.g., Pysek and Liska 1991; Totland, Grytnes, and Heegaard 2004) or higher species richness outside the nurses (e.g., Cavieres et al. 1998). To illustrate this disparity of effects, we summarized our findings on the effects of alpine cushion plant species on species richness at different latitudes along the Andes of southern South America (see Cavieres et al. 1998, 2002, 2005, 2006; Badano et al. 2002; Arroyo et al. 2003; Badano and Cavieres 2006a, 2006b). Species richness was never lower within cushions than in open areas away from cushions, but not all cushions contained more species than the surrounding habitat (Figure 3.1). Positive effects of cushion on species richness were detected at 30° S, 41° S, and 50° S, where

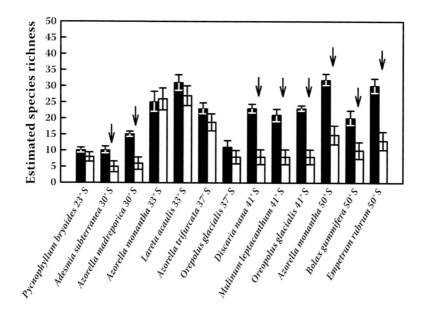

FIGURE 3.1 Comparisons of species richness between cushions (solid bars) and open areas (empty bars) at different sites along the Andes of southern South America. The x-axis indicates the latitude of the study site and the name of the cushion species dominating the plant community under study. Error bars are 95% confidence intervals, and arrows indicate significant differences (nonoverlapped confidence intervals) between cushions and open areas.

the number of species within cushions was 2.0–2.9 times higher than outside them. However, no differences in species richness were detected between habitats at 23° S, 33° S, and 37° S. These results suggest that patch-scale effects of cushion on species richness are rather idiosyncratic across latitudes. Nonetheless, it is important to note that cushions within the same latitude showed similar effects, either positive or neutral (see Figure 3.1), suggesting that the effect of cushions on species richness may be related to the environmental conditions (see Section 3.2.4).

Therefore, as in arid and semi-arid habitats, no clear trends regarding species richness emerge from comparisons at the patch level, and there is no unequivocal support for the idea that species richness is higher within patches created by nurses, suggesting idiosyncratic effects of nurses (Badano and Cavieres 2006a). However, as we will see below, many of these idiosyncratic responses depend on the intensity of abiotic stress experienced by the plants in the microhabitats away from nurses, and on the ability of nurses to mitigate such stressful conditions.

3.2.2 Diversity and Species Abundance Relations

Although species richness provides an important basis for comparisons of the effects of nurses within the community, it says nothing about the relative commonness and rarity of species. Calculation of diversity and evenness indexes and graphical representations of the species abundance rankings could help in the visualizations of the changes induced by the presence of facilitator species.

For instance, Rebollo et al. (2002) studied the protective role against herbivores of the cactus species *Opuntia polyacantha* in shortgrass steppes in North America, and found that species diversity measured by the Shannon-Weaver index (H′) was higher beneath the nurses than outside them. Species richness and composition were the same for both habitats, indicating that differences in diversity were due to a lower dominance (i.e., a more-even distribution of dominance) in patches beneath the nurses than outside them. Callaway, Kikvidze, and Kikodze (2000) evaluated the effects of two unpalatable weeds (*Cirsium obvalatum* and *Veratrum lobelianum*) in protecting palatable species from grazing by cattle in the Caucasus Mountains. Using species-rank abundance curves, they showed that rare species in open meadow areas were more abundant when associated with the unpalatable plants, indicating a more even abundance distribution within the facilitator species than outside them. Likewise, Pysek and Liska (1991), studying the association with the cushion plant *Sibbaldia tetrandra* in the alpine zone of Pamiro-Alai, Russia, reported that the plant assemblages growing within cushions had a more even distribution of abundance than the assemblage growing outside cushions, despite no differences in the overall species richness. In contrast, Badano et al. (2002) reported that the nurse cushion species *Oreopulus glacialis* decreased diversity in the high Andes of central Chile by favoring a greater abundance of a few species, thereby decreasing the evenness in the species abundance (see Figure 3.2). In contrast, cushions of *Pycnophyllum bryoides* generated species assemblages with a more even distribution of species abundance than what occurred in open areas (Figure 3.2).

These examples illustrate the importance that facilitation by nurse species can have in changing the rarity and commonness of some particular species, which may have important implications for the maintenance and functioning of the whole community.

3.2.3 SPECIES COMPOSITION

As we have seen, the amelioration of abiotic environments does not necessarily affect the number of species, but it could affect the abundance of some particular species, generating changes in the structure of the species assemblage growing beneath nurses compared with that developed outside them. In addition, the different microclimates generated by the nurse species can produce conditions such that the species able to grow beneath nurses are completely different from those able to grow outside nurses. Ordination analyses are among the simplest ways to assess these patch-scale differences in species composition between nurse-dominated patches and their surrounding habitat.

For example, in the study of Callaway, Kikvidze, and Kikodze (2000), detrended correspondence analysis (DCA) ordinations showed substantial differences in species composition between areas beneath nurses and areas away from them, where these differences were mainly due to a higher association of palatable species with the unpalatable weeds. Tewksbury and Lloyd (2001) also used DCA ordinations to analyze the consequences at large spatial scales of the nurse effects of *Olneya tesota* trees in the Sonoran Desert. At a landscape scale, which comprised a mixture of

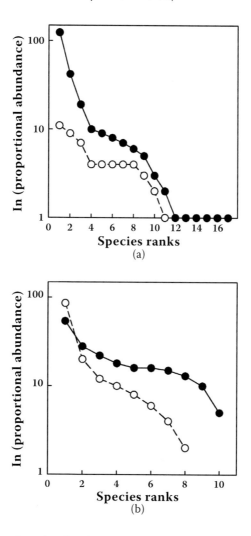

FIGURE 3.2 Ranks of species abundance in two alpine communities of the high Andes of Chile. (a) Comparison of the species rank abundance within (black circles) and outside (empty circles) for *Oreopolus glacialis* cushions. (b) Comparison of the species rank abundance within (black circles) and outside (empty circles) for *Pycnophyllum bryoides* cushions.

samples taken in mesic and xeric sites, these authors found that the presence of the nurse did not explain the variation in community structure. However, when ordinations were performed separately for xeric and mesic sites, they noted that differences in composition of species assemblages occurred more frequently in xeric than in mesic sites, suggesting that the facilitative effects of *Olneya* canopies and the consequences for differentiation in the species assemblages were more important in the harsher sites (see also Suzán, Nabhan, and Patten 1996).

Studies conducted in alpine areas have also shown that nurse species have important effects on the structure of species assemblages. For example, Totland, Grytnes, and Heegaard (2004) reported consistent differences (via DCA ordinations) in the species assemblages associated with *Salix lapponum* compared with open areas in different alpine sites of Norway. Pysek and Liska (1991) also reported that the species composition and abundance—growing within the cushion plant *Sibbaldia tetrandra* in the alpine zone of Pamiro-Alai, Russia—are different from those found in areas away from cushions. Nonmetric multidimensional scaling (NMS) ordinations performed with our data set, collected while evaluating the importance of the nurse effect of alpine cushion species along the Andes of southern South America, showed consistent difference between the species assemblages developed within cushions compared with those found outside cushions (Figure 3.3). The single exception is the community dominated by *Azorella trifurcata* cushions at 37° S, which is located in the foothills of an active volcano (Antuco Volcano).

These results highlight that nurses harbor species that are not present in open areas and vice versa, and/or they generate different species-abundance relations compared with open areas. These conclusions have important consequences for diversity at the entire community level (see Section 3.3).

3.2.4 EFFECTS OF NURSE SIZE: WHEN SIZE DOES MATTER

The magnitude of the microenvironmental changes generated by the presence of a nurse species is highly related to its size or age (Pugnaire et al. 1996), and it has been suggested that the importance of the facilitative effect of a nurse is directly proportional to its size (Callaway and Walker 1997).

Tewksbury and Lloyd (2001) reported that nurse size has significant effects on the number of species harbored by *Olneya tesota*. However, this was true only for perennial species; richness of ephemeral species was not affected by the size of the nurse. Likewise, Maestre and Cortina (2005) analyzed the importance of different shrub species on species richness in their understory in a semi-arid zone in Spain and found significant effects of the shrub size on species richness underneath. Interestingly, the relationship between shrub size and underneath species richness was not affected by the identity of the shrub species, indicating equivalence on this effect for the different nurse species. Pugnaire et al. (1996) reported that the age of the nurse shrub *Retama sphaerocarpa* is correlated with the size of its canopy, and showed that the number of plant species and abundance of seedlings increased with the age of the nurse, hence with its size (see also Pugnaire and Lázaro 2000).

Studies conducted in alpine habitats have also found strong positive relationships between the size of the nurse and the species richness harbored (e.g., Pysek and Liska 1991; Nuñez, Aizen, and Ezcurra 1999; Cavieres et al. 2002; Arroyo et al. 2003). To assess the consistency of this effect, we performed linear regression (log–log) between the number of species growing within nurse cushion species and cushion size, and compared these relationships with those obtained for the equivalent open areas away from cushions at different location along the southern Andes of South America. As expected, species richness within cushions, as well as in surrounding open areas, increased significantly with area (Figure 3.4). However, intercepts of regression functions for cushions were always higher than those obtained for open areas ($p < .01$ in all cases), indicating that cushion patches support a higher species

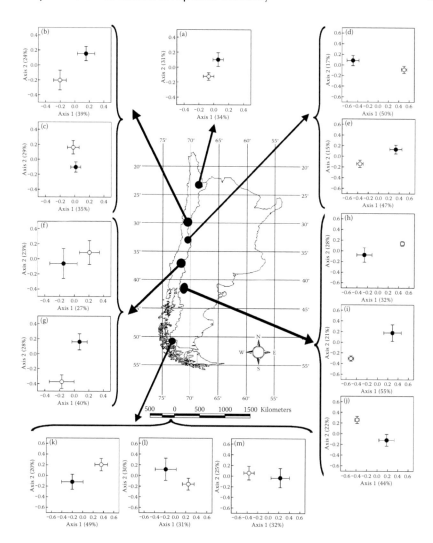

FIGURE 3.3 Results of NMS ordinations comparing species composition between cushions (solid symbols) and open areas (empty symbols) at the different sites along the Andes of southern South America. Symbols are centroids, and bidirectional error bars are the 95% confidence intervals. Nonoverlapped bidirectional confidence intervals indicate significant differences between cushions and open areas either on the first or second ordination axis. (a) *Pycnophyllum bryoides*, 23° S; (b) *Adesmia subterranea*, 30° S; (c) *Azorella madreporica*, 30° S; (d) *Azorella monantha*, 33° S; (e) *Laretia acaulis*, 33° S; (f) *Azorella trifurcata*, 37° S; (g) *Oreopolus glacialis*, 37° S; (h) *Discaria nana*, 41° S; (i) *Mulinum leptacanthum*, 41° S; (j) *Oreopolus glacialis*, 41° S; (k) *Azorella monantha*, 50° S; (l) *Bolax gummifera*, 50° S; and (m) *Empetrum rubrum*, 50° S.

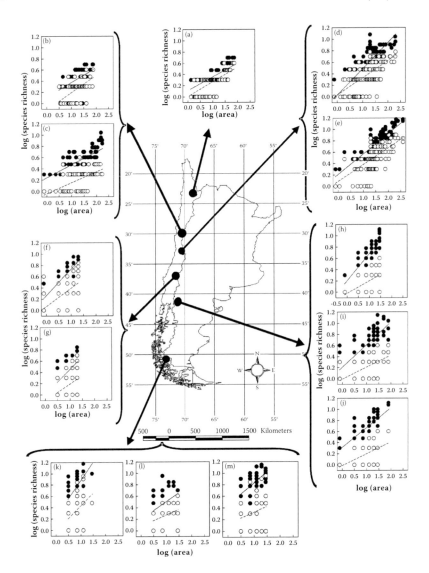

FIGURE 3.4 Relationships between the number of species and the area of samples within (solid symbols–solid lines) and outside cushions (empty symbols–dashed lines) at the different sites considered for the analyses. (a) *Pycnophyllum bryoides*, 23° S; (b) *Adesmia subterranea*, 30° S; (c) *Azorella madreporica*, 30° S; (d) *Azorella monantha*, 33° S; (e) *Laretia acaulis*, 33° S; (f) *Azorella trifurcata*, 37° S; (g) *Oreopolus glacialis*, 37° S; (h) *Discaria nana*, 41° S; (i) *Mulinum leptacanthum*, 41° S; (j) *Oreopolus glacialis*, 41° S; (k) *Azorella monantha*, 50° S; (l) *Bolax gummifera*, 50° S; and (m) *Empetrum rubrum*, 50° S. In all cases, we used log–log transformations to reach normality of data.

density than open areas, even at the smallest sample size. While slopes of regression functions showed no differences between cushions and open areas at latitudes lower than 33° S ($p > .05$), in latitudes higher than 33° S slopes of regressions for cushions were significantly higher than those obtained for open areas ($p < .01$). These results indicate that in southern latitudes (>33° S), species accumulation rate within cushions is higher than that in open areas, suggesting a positive synergistic effect of nurse size on species richness.

As stated, the size of the nurse is usually correlated with the magnitude of the microclimatic differences with open areas generated by the nurse (Pugnaire et al. 1996; Callaway and Walker 1997). Nonetheless, the fact that size and age are also usually correlated suggests that the positive relationship between nurse size and species richness is a mixture of the mitigation of harsh conditions found in the open areas and the time that nurses have been exposed for colonization. Comparisons of the species–area relationship obtained for both the nurse plant and the open areas away from the nurse could shed light about the relative importance of both processes. For instance, the fact that the species accumulation rate on cushion plants located in the extremely harsh alpine zone of the Patagonian Andes (latitude 50° S) was higher than the accumulation in open areas suggests that microclimatic ameliorations play a more important role than the time available for colonization (Cavieres et al. 2002; Arroyo et al. 2003).

3.2.5 Changes along Severity Gradients

Following Bertness and Callaway's (1994) model, which predicted higher importance of positive interactions as the environmental harshness increases, Hacker and Gaines (1997) suggested that the impact of a facilitator species on the enhancement of species richness will be higher as the environment become more harsh (see also Michalet et al. 2006). Hence, it could be expected that along an environmental severity gradient, the differences in the number of species harbored by nurses in comparison with the open areas away from them will be higher as the abiotic environment become harsher.

Tewksbury and Lloyd (2001) visited different locations along the Sonoran Desert that differed in water availability, and found that on mesic sites, while species richness of perennial plants beneath ironwood trees (*Olneya tesota*) did not differ from that found outside the nurse, the richness of ephemeral species was higher outside nurses. On xeric sites, however, while species richness of ephemeral plants did not differ between underneath nurses and outside them, richness of perennial species was higher underneath nurses. Similar results have been recently reported by Holzapfel et al. (2006), who studied an extensive aridity gradient from mesic Mediterranean sites to an arid site in Israel. They found that while species density (species per area unit) in mesic sites did not differ between underneath shrubs and open areas, species density in the shrubs underneath was twice as high in the arid sites.

Studies conducted at two contrasting elevations in the Chilean Patagonian Andes showed that, while alpine cushion species did not have significantly higher species richness than open areas away from them at lower elevations, at higher elevations, cushions contained three- to fourfold more species than outside them (Cavieres et al. 2002; Arroyo et al. 2003). Similar effects have also been reported for the Andes of central Chile (30°–33° S; Badano and Cavieres 2006b), where the magnitude of the positive effect of three cushion species (*Azorella monantha*, *A. madreporica*, and *Adesmia subterranea*) on species richness increased with elevation.

These findings demonstrate that the impact of nurse plants on species richness depends on the importance of its facilitative effect, which is related to the severity of the abiotic environment, and on the characteristics of the species involved. From this, it seems clear that the importance of a nurse species in maintaining the diversity of a community will be higher as environments become more stressful. Interestingly, these effects emerge more clearly in studies that evaluated wide geographical gradients (e.g., Holzapfel et al. 2006), where the fact that both environmental conditions and plant community change enable a greater clarification of the key processes that drive community assembly.

3.3   EFFECTS AT THE ENTIRE COMMUNITY LEVEL

Species assemblages growing within and outside nurses are parts of the same community. Thus, to assess the effects of the presence of facilitator species on richness at the entire community level, we must consider at the same time both the habitat patches created by nurses and the surrounding environment without nurses, and then compare that with an ideal situation without the presence of facilitator species. In this section, we analyze the consequences of nurses for species richness at the entire community level.

We have seen that the effects on species richness at the patch scale are rather idiosyncratic, and depend on the severity of the surrounding environments and on the intensity of the amelioration performed by the nurse. Nonetheless, species composition analyses systematically suggest that nurses harbor different species than those found outside them. Therefore, nurses might increase the overall species richness of the entire community.

De Villiers, Van Rooyen, and Theron (2001) studied seedling emergence and survival beneath and between the canopy of five shrub species in the Strandveld Succulent Karoo in South Africa, and found that species richness and seedling densities were significantly higher in open areas than underneath shrubs. Hence, these authors concluded that no evidence of facilitation was found on that site. Nevertheless, a thorough revision of their results showed that 15 species only grew beneath shrubs, indicating that ca. 23% of the local species richness at that site is due to the presence of shrubs. Interestingly, most of the species that were restricted to areas beneath shrubs were perennial herbaceous, whereas species growing in open areas away from shrubs were ephemeral (annual) species. In a similar study conducted in an arid area of northern Chile, Gutiérrez et al. (1993) reported 19 species growing beneath the canopy of the shrub *Porlieria chilensis*, whereas 29 species were found in open areas between shrubs. Nonetheless, despite the ca. fourfold higher density of individuals outside shrubs, no differences in total biomass were found between these two microhabitats, indicating that the fewer individuals growing beneath shrubs attained higher biomass. Interestingly, six species (ca. 17% of the local richness) were found only beneath shrubs. Likewise, Pugnaire, Armas, and Valladares (2004) reported that shrub patches do not contain more species than patches of similar size in open areas away from them; they found 14 species (ca. 40% of the total species reported for that site) exclusively growing beneath shrubs.

Studies conducted on alpine habitats also demonstrate the impact of nurse species on species richness of the entire community. For example, in the Patagonian Andes of Argentina, Nuñez, Aizen, and Ezcurra (1999) found that patches (20 × 20 m) where nurse cushion species are present contained higher species richness than patches where cushions are absent (see also Badano and Cavieres 2006a, 2006b).

If $D_{-\text{nurses}}$ is the diversity that a given community has in the absence of nurses (i.e., diversity in open areas) and $D_{+\text{nurses}}$ is the diversity that this community actually has because of the presence of nurses, we can estimate the magnitude of the enhancement on species richness in the community by modifying the index RII (relative interaction index) proposed by Armas, Ordiales, and Pugnaire (2004). Thus, the magnitude of the positive effect (MPE) on species richness due to the nurses can be assessed as follows:

$$\text{MPE} = \frac{(D_{+\text{nurses}} - D_{-\text{nurses}})}{(D_{+\text{nurses}} + D_{-\text{nurses}})}$$

where MPE can range from 0 (no effects) to 1 (all species depend on the presence of nurses and, hence, all of the diversity of the community). If we multiply this index by 100, then this expression refers to the percentage of species within the community added due to the presence of nurses. We evaluated the effects of cushion plants on species richness of the entire community, where we used data from open areas as an estimation of the species diversity that communities would have had in the absence of nurses (i.e., $D_{-\text{nurses}}$ = species diversity in open areas). Data from cushions and open areas were pooled in a single data set to estimate the actual species diversity of the entire community at each site ($D_{+\text{nurses}}$ = species diversity in open areas plus cushions patches).

MPE values indicated a strong variation in the effects of cushions on species richness at the entire community level (Table 3.1). With the exception of *Azorella trifurcata* at 37° S, substantial increases in species richness due to the presence of cushions were indicated at all sites along the high Andes (Figure 3.5). The strongest effect was detected for *Mulinum leptacanthum* at 41° S, where this cushion was indicated to support up to 52% of species within the community. Although this kind of effect has not been previously reported for other nurse species, a few published studies that have reported the number of species within and outside nurse-dominated patches allow us to calculate MPE and provide other examples of these effects. In the previous example from the Caucasus Mountains, where unpalatable weeds protected palatable species from grazing (Callaway, Kikvidze, and Kikodze 2000), at least two species (out of 34) were detected only within these weed patches, which report a low MPE on species richness (6%). In the community survey evaluating the nurse effects of *Olneya tesota* trees in the Sonoran Desert, Suzán, Nabhan, and Patten (1996) reported a total of 88 species, including trees, shrubs, and cacti, where 12 of these species were found only beneath *O. tesota* shade, indicating an MPE of 14%.

TABLE 3.1

Summary of Landscape-Scale Effects of Cushion Plants Being Present at Different Latitudes along the High Andes

Cushion Species	Latitude	Species Level	Genus Level	Family Level	Endemics
Pycnophyllum bryoides	23° S	16% (11)	21% (9)	12% (5)	34% (2)
Adesmia subterranea	30° S	33% (12)	25% (10)	45% (8)	0% (4)
Azorella madreporica	30° S	46% (16)	33% (12)	50% (9)	20% (6)
Azorella monantha	33° S	12% (35)	9% (29)	13% (19)	14% (17)
Laretia acaulis	33° S	16% (39)	17% (28)	10% (17)	27% (19)
Azorella trifurcate	37° S	7% (26)	4% (22)	4% (14)	28% (4)
Oreopolus glacialis	37° S	22% (17)	21% (15)	16% (11)	11% (2)
Mulinum leptacanthum	41° S	52% (25)	49% (23)	41% (19)	45% (8)
Discaria nana	41° S	51% (25)	49% (23)	38% (18)	40% (7)
Oreopolus glacialis	41° S	51% (24)	49% (23)	38% (18)	45% (8)
Azorella monantha	50° S	33% (34)	9% (25)	28% (15)	28% (8)
Bolax gummifera	50° S	38% (27)	29% (21)	22% (12)	29% (8)
Empetrum rubrum	50° S	35% (35)	22% (24)	24% (14)	21% (8)

Note: Values are percentages of species, genera, families, and endemics whose presence in communities depends on cushion plants. The total numbers of species, genera, families, and endemics for each community are indicated in parentheses.

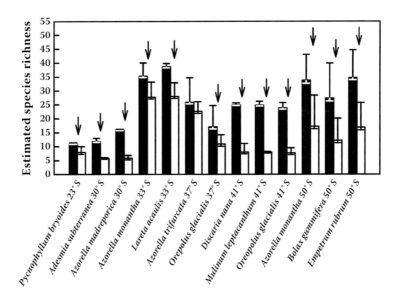

FIGURE 3.5 Estimated species richness, including data from landscapes composed of open areas plus cushions (solid bars) and open areas only (empty bars). The x-axis indicates the latitude of the study site and the name of the cushion species dominating the plant community under study. Arrows indicate significant differences (nonoverlapped confidence intervals) between landscapes with and without cushions.

These examples illustrate that, despite some studies reporting higher species richness in areas outside nurses, a careful revision of their data indicate that there are species restricted to the nurse microhabitat and, hence, without the presence of nurses, these species can no longer persist in the community. Therefore, as long as the persistence of a particular group of species in a given community depends on the presence of a nurse species, the effect of the facilitation by nurses will increase species richness at the local scale.

3.4 TAKING A LOOK BEYOND THE SPECIES LEVEL

Species that depend on the presence of a nurse species for their persistence in a community could belong to different genera and families than those able to grow in open areas away from nurses. Thus, facilitation by nurses could increase the taxonomic richness of a community beyond the species level (i.e., genera, families), enabling the existence of taxa with different historical and evolutionary trajectories. This suggests that facilitation could permit the existence of lineages that otherwise could not be present in the community, which has important implications for the maintenance of diversity from a historical perspective (see Valiente-Banuet et al. 2006; Valiente-Banuet and Verdú 2007).

By focusing on the effects of cushion plants in high-Andean landscapes, we asked whether the presence of cushions leads to increased diversity of genera and families. To estimate the effects of cushion plants on the number of genera and families, we used an approach similar to the one outlined in Section 3.3 by estimating the MPE of cushions at these taxonomic levels. Although the magnitude of the MPE effect at the genera and family levels showed high variation between sites and cushion species, in all cases cushions increased the number of genera and families within communities (Table 3.1). The higher positive effects at the genus level were obtained for the cushions at 41° S, where the three cushion species from this latitude (*Mulinum leptacanthum, Discaria nana*, and *Oreopolus glacialis*) were indicated to support up to 49% of the genera present in that community. On the other hand, the lowest effect was detected at 37° S, where *Azorella trifurcata* cushions contributed only 4% to the number of genera. At the family level, the highest positive effect was detected at 30° S, where 50% of the families within that community depended on the presence of *Azorella madreporica* cushions. Similar to the effects at the genus level, the lowest positive effect at the family level was detected for *Azorella trifurcata* at 37° S.

Finally, endemic species are considered to be the representation of the uniqueness of a site or biota, and therefore are particularly important for conservation issues. If nurse species harbor more endemic species than open areas, then conservation efforts should be focused on these key elements for the maintenance of the unique biodiversity of a site or region. We evaluated whether the presence of cushions was supporting the presence of endemic species in the Andean landscapes, and found that the magnitude of the effects of cushions on this group of species ranged from null at 30° S, where cushions of *Adesmia subterranea* had no effects on the presence of endemic species, to highly positive at 41° S, where cushion plants maintain 40%–45% of the endemic species present in that community (Table 3.1). Therefore, in those habitats where nurse species harbor a substantial proportion of the endemic species, the conservation of the nurses seems to provide a new venue for effective conservation of endemics at the whole-community level (see also Padilla and Pugnaire 2006).

3.5 CONCLUDING REMARKS

Community ecologists recognize that many factors affect the species composition of a given community, with no single factor providing a complete explanation for observed patterns (Lortie et al. 2004). Moreover, different factors can interact in a complex hierarchical fashion. The pool of regional species provides the potential members of a community; the dispersal ability sets the identity of those species available to colonize a given community, where interspecific interactions will play a fundamental role in the success or failure of species as community members.

We have seen that the presence of facilitator species (i.e., species that mitigate mortality factors) generates a variety of changes in different community attributes, where the most relevant result is the enhancement of species richness at the whole-community level. Most importantly, facilitation of richness enhancement enables the coexistence of taxa with different historical and evolutionary trajectories, decreasing the chances of redundancy in the function of an ecosystem. Communities with greater diversity may have several enhanced processes relevant for the function of an ecosystem (e.g., carbon and nitrogen fixation); hence, the role of facilitation on ecosystem functioning and on the balance of nature seems to be very relevant (see Callaway 2007). Environments in which facilitation is a key process (e.g., alpine and arid systems) are particularly sensitive to major anthropogenic drivers of ecosystem change, including climate change, land-use change, and introduced species. Thus, for proper management and conservation of these environments, it is vital to understand the importance of facilitation in the regulation of their biodiversity.

ACKNOWLEDGMENTS

We thank Maritza Mihoc, Angela Sierra, Susana Gómez-González, Marco Molina-Montenegro, and Cristián Torres-Diaz for their help on the field work and on different stages of this research. This study was supported by FONDECYT 1030821 and 1060710, and by P05-002 F ICM and PFB-23 supporting the Center for Advanced Studies in Ecology and Research on Biodiversity (IEB).

REFERENCES

Armas, C., R. Ordiales, and F. I. Pugnaire. 2004. Measuring plant interactions: A new comparative index. *Ecology* 85: 2682–2686.
Arroyo, M. T. K., L. A. Cavieres, A. Peñaloza, and M. A. Arroyo-Kalin. 2003. Positive associations between the cushion plant *Azorella monantha* (Apiaceae) and alpine plant species in the Chilean Patagonian Andes. *Plant Ecology* 169: 121–129.
Badano, E. I., and L. A. Cavieres. 2006a. Ecosystem engineering across ecosystems: Do engineer species sharing common features have generalized or idiosyncratic effects on species diversity? *Journal of Biogeography* 33: 304–313.

Badano, E. I., and L. A. Cavieres. 2006b. Impacts of ecosystem engineers on community attributes: Effects of cushion plants at different elevations of the Chilean Andes. *Diversity and Distribution* 12: 388–396.

Badano, E. I., L. A. Cavieres, M. Molina-Montenegro, and C. Quiroz. 2005. Slope aspect influences plant association patterns in the Mediterranean matorral of central Chile. *Journal of Arid Environment* 62: 93–108.

Badano, E. I., C. G. Jones, L. A. Cavieres, and J. P. Wright. 2006. Assessing impacts of ecosystem engineers on community organization: A general approach illustrated by effects of a high-Andean cushion plant. *Oikos* 115: 369–385.

Badano, E. I., M. A. Molina-Montenegro, C. Quiroz, and L. A. Cavieres. 2002. Efectos de la planta en cojín *Oreopolus glacialis* (Rubiaceae) sobre la riqueza y diversidad de especies en una comunidad alto- andina de Chile central. *Revista Chilena de Historia Natural* 75: 757–765.

Bertness, M. D., and R. M. Callaway. 1994. Positive interactions in communities. *Trends in Ecology and Evolution* 9: 191–193.

Brooker, R. W., and T. V. Callaghan. 1998. The balance between positive and negative interactions and its relationship to environmental gradient: A model. *Oikos* 81: 196–207.

Brooker, R. W., F. T. Maestre, R. M. Callaway, C. L. Lortie, L. A. Cavieres, G. Kunstler, P. Liancourt, et al. 2008. Facilitation in plant communities: The past, the present and the future. *Journal of Ecology* 96: 18–34.

Bruno, J. F., J. J. Stachowicz, and M. D. Bertness. 2003. Inclusion of facilitation into ecological theory. *Trends in Ecology and Evolution* 18: 119–125.

Callaway, R. M. 1995. Positive interactions among plants. *Botanical Review* 61: 306–349.

Callaway, R. M. 2007. *Positive interactions and interdependence in plant communities.* Berlin: Springer.

Callaway, R. M., R. W. Brooker, P. Choler, Z. Kikvidze, C. J. Lortie, R. Michalet, L. Paolini, et al. 2002. Positive interactions among alpine plants increase with stress. *Nature* 417: 844–848.

Callaway, R. M., Z. Kikvidze, and D. Kikodze. 2000. Facilitation by unpalatable weeds may conserve plant diversity in overgrazed meadows in the Caucasus Mountains. *Oikos* 89: 275–282.

Callaway, R. M., and F. I. Pugnaire. 1999. Facilitation in plant communities. In *Handbook of functional plant ecology*, ed. F. Pugnaire and F. Valladares, 623–648. New York: Dekker.

Callaway, R. M., and L. R. Walker. 1997. Competition and facilitation: A synthetic approach to interactions in plants communities. *Ecology* 78: 1958–1965.

Cavieres, L., M. T. K. Arroyo, A. Peñaloza, M. Molina-Montenegro, and C. Torres. 2002. Nurse effect of *Bolax gummifera* cushion plants in the alpine vegetation of the Chilean Patagonian Andes. *Journal of Vegetation Science* 13: 547–554.

Cavieres, L. A., E. I. Badano, A. Sierra-Almeida, S. Gómez-González, and M. Molina-Montenegro. 2006. Positive interactions between alpine plant species and the nurse cushion plant *Laretia acaulis* do not increase with elevation in the Andes of central Chile. *New Phytologist* 169: 59–69.

Cavieres, L. A., A. P. G. Peñaloza, C. Papic, and M. Tambutti. 1998. Efecto nodriza de *Laretia acaulis* en plantas de la zona andina de Chile central. *Revista Chilena de Historia Natural* 71: 337–347.

Cavieres, L. A., C. Quiroz, M. Molina-Montenegro, A. Pauchard, and M. A. Muñoz. 2005. Nurse effect of the native cushion plant *Azorella monantha* on the invasive non-native *Taraxacum officinale* in the high-Andes of central Chile. *Perspectives in Plant Ecology, Systematics and Evolution* 7: 217–226.

Choler P., R. Michalet, and R. M. Callaway. 2001. Facilitation and competition on gradients in alpine plant communities. *Ecology* 82: 3295–3308.

Connell, J. H. 1978. Diversity in tropical rain forests and coral reefs. *Science* 199: 1302–1310.

de Villiers, A. J., M. W. Van Rooyen, and G. K. Theron. 2001. The role of facilitation in seedling recruitment and survival patterns, in the Strandveld Succulent Karoo, South Africa. *Journal of Arid Environments* 49: 809–821.

Fuentes, E. R., R. D. Otaiza, M. C. Alliende, A. Hoffmann, and A. Poiani. 1984. Shrub clamps of the Chilean matorral vegetation: Structure and possible maintenance mechanisms. *Oecologia* 62: 405–411.

Grime, J. P. 1973. Competitive exclusion in herbaceous vegetation. *Nature* 242: 344–347.

Gutiérrez, J. R., P. L. Meserve, L. C. Contreras, H. Vásquez, and F. M. Jaksic. 1993. Spatial distribution of soil nutrients and ephemeral plants underneath and outside the canopy of *Porlieria chilensis* shrubs (Zygophyllaceae) in arid coastal Chile. *Oecologia* 95: 347–352.

Hacker, S. D., and S. D. Gaines. 1997. Some implications of direct positive interactions for community species diversity. *Ecology* 78: 1990–2003.

Holzapfel, C., K. Tielbörger, H. A. Parag, J. Rigel, and M. Sternberg. 2006. Annual plant-shrub interactions along an aridity gradient. *Basic and Applied Ecology* 7: 268–279.

Jones, C. G., J. H. Lawton, and M. Shachak. 1997. Positive and negative effects of organisms as physical ecosystem engineers. *Ecology* 78: 1946–1957.

Keeley, S. C., and A. W. Johnson. 1977. A comparison of the patterns of herb and shrub growth in comparable sites of Chile and California. *American Midland Naturalist* 97: 120–132.

Kikvidze, Z., F. I. Pugnaire, R. W. Brooker, P. Choler, C. J. Lortie, R. Michalet, and R. M. Callaway. 2005. Linking patterns and processes in alpine plant communities: A global study. *Ecology* 86: 1395–1400.

Larrea-Alcázar, D. M., R. P. López, and D. Barrientos. 2005. The nurse-plant effect of *Prosopis flexuosa* D.C. in a dry valley of the Bolivian Andes. *Ecotropicos* 16: 89–95.

Lortie, C. J., R. W. Brooker, P. Choler, Z. Kikvidze, R. Michalet, F. I. Pugnaire, and R. M. Callaway. 2004. Rethinking plant community theory. *Oikos* 107: 433–438.

Maestre, F. T., S. Bautista, and J. Cortina. 2003. Positive, negative and net effects in grass-shrub interactions in Mediterranean semi-arid grasslands. *Ecology* 84: 3186–3197.

Maestre, F. T., and J. Cortina. 2005. Remnant shrubs in Mediterranean semi-arid steppes: Effects of shrub size, abiotic factors, and species identity on understorey richness and occurrence. *Acta Oecologica* 27: 161–169.

Michalet, R. 2006. Is facilitation in arid environments the result of direct or complex interactions? *New Phytologist* 169: 3–6.

Michalet, R. 2007. Highlighting the multiple drivers of change in interactions along stress gradients. *New Phytologist* 173: 3–6.

Michalet, R., R. W. Brooker, L. A. Cavieres, Z. Kikvidze, C. J. Lortie, F. I. Pugnaire, A. Valiente-Banuet, and R. M. Callaway. 2006. Do biotic interactions shape both sides of the humped-back model of species richness in plant communities? *Ecology Letters* 9: 767–773.

Morin, P. J. 1999. *Community Ecology*. Malden, MA: Blackwell Science.

Nuñez, C. I., M. A. Aizen, and C. Ezcurra. 1999. Species associations and nurse plant effects in patches of high-Andean vegetation. *Journal of Vegetation Science* 10: 357–364.

Padilla, F. M., and F. I. Pugnaire. 2006. The role of nurse plant in the restoration of degraded environments. *Frontiers in Ecology and the Environment* 4: 196–202.

Powers, J. S., J. P. Haggar, and R. F. Fisher. 1997. The effect of overstory composition on understory woody regeneration and species richness in a 7-years-old plantation in Costa Rica. *Forest Ecology and Management* 99: 43–54.

Pugnaire, F. I., C. Armas, and F. Valladares. 2004. Soil as a mediator in plant-plant interactions in a semi-arid community. *Journal of Vegetation Science* 25: 85–92.

Pugnaire, F. I., P. Haase, J. Puigdefábregas, M. Cueto, S. C. Clark, and D. Incoll. 1996. Facilitation and succession under canopy of a leguminous shrub, *Retama sphaerocarpa*, in a semi-arid environment in south-east Spain. *Oikos* 76: 455–464.

Pugnaire, F. I., and R. Lázaro. 2000. Seed bank understorey species composition in a semi-arid environment: The effect of shrub age and rainfall. *Annals of Botany* 86: 807–813.

Pysek, P., and J. Liska. 1991. Colonization of *Sibbaldia tetrandra* cushions on alpine scree in the Pamiro-Alai Mountains, Central Asia. *Arctic and Alpine Research* 23: 263–272.

Raffaele, E., and T. Veblen. 1998. Facilitation by nurse shrubs of resprouting behavior in a post-fire shrubland in northern Patagonia, Argentina. *Journal of Vegetation Science* 9: 693–698.

Rebollo, S., D. G. Milchunas, I. Noy-Meir, and P. L. Chapman. 2002. The role of a spiny plant refuge in structuring grazed shortgrass steppe plant communities. *Oikos* 98: 53–64.

Rossi, B. E., and P. E. Villagra. 2003. Effects of *Prosopis flexuosa* on soil properties and the spatial pattern of understorey species in arid Argentina. *Journal of Vegetation Science* 14: 543–550.

Suzán, H., G. P. Nabhan, and D. T. Patten. 1996. The importance of *Olneya testosa* as a nurse plant in the Sonoran Desert. *Journal of Vegetation Science* 7: 635–644.

Tilman, D. 1982. *Resource competition and community structure*. Princeton, NJ: Princeton University Press.

Tewksbury, J. J., and J. D. Lloyd. 2001. Positive interactions under nurse-plants: Spatial scale, stress gradients and benefactor size. *Oecologia* 127: 425–434.

Totland, O., J. A. Grytnes, and E. Heegaard. 2004. Willow canopies and plant community structure along an alpine environmental gradient. *Arctic, Antarctic, and Alpine Research* 36: 428–435.

Valiente-Banuet, A., and M. Verdú. 2007. Facilitation can increase the phylogenetic diversity of plant communities. *Ecology Letters* 10: 1029–1036.

Valiente-Banuet, A., A. Vital, M. Verdú, and R. M. Callaway. 2006. Modern quaternary plant lineages promote diversity through facilitation of ancient tertiary lineages. *Proceedings of the National Academy of Science* 103: 16812–16817.

4 Biotic Interactions, Biodiversity, and Community Productivity

Richard Michalet and Blaise Touzard

CONTENTS

4.1 INTRODUCTION

The global decline in biodiversity induced by human activities has prompted scientists to investigate the potential value of biodiversity for ecosystem services (Chapin et al. 2000; Schröter et al. 2005; Diaz et al. 2007). A large number of experiments have been conducted in the last decade, in particular in herbaceous communities, in order to assess the potential roles of different components of biodiversity for community productivity, stability, and invasibility (see Hooper et al. 2005). Because most results of these experiments were considered to be opposite to the natural patterns of biodiversity along productivity gradients (Loreau et al. 2001), this fueled substantial debate about the underlying mechanisms of the effect of biodiversity on ecosystem functions (e.g., Huston et al. 2000). Furthermore, this created a new interest in the ecological drivers of biodiversity in natural environments, a topic of research that had interested plant ecologists in the past (Whittaker 1972; Grime 1973; Connell 1978; Huston 1979).

In theoretical ecology, biotic interactions are considered to be important drivers of community composition and richness, together with chance biogeographical events (e.g., dispersal) and local environmental factors (Lortie et al. 2004; see also Figure 6.2 in Chapter 6, this volume). All hypotheses addressing the role of biodiversity for ecosystem functioning have also emphasized the crucial importance of biotic interactions. In this chapter we focus on (a) the role of negative and positive interactions (competition and facilitation) in driving natural patterns of biodiversity (Section 4.2) and (b) the effect of biodiversity for ecosystem functions and in particular productivity (Section 4.3). We will detail the main theoretical models proposed in the literature, while other chapters of this book will present more detailed results on both aspects of the relationship between diversity and productivity. In Section 4.4, we focus on the debate about the discrepancy between natural patterns of biodiversity and the results of diversity experiments. We will propose alternative reconciliations to those existing in the literature, in particular two hypotheses emphasizing the roles of facilitation and niche complementarity for both biodiversity and productivity along natural environmental gradients and in diversity experiments.

4.2   BIOTIC INTERACTIONS AND BIODIVERSITY

4.2.1   COMPETITION AND BIODIVERSITY

One important initial step for understanding the role of biotic interactions for biodiversity was done through the development and refining of the niche concept and in particular with the proposition of Hutchinson (1959) to distinguish the fundamental from the realized niche (Bruno, Stachowicz, and Bertness 2003; see Figure 4.1a). At this time, only competition was thought to affect the spatial distribution of species along environmental gradients and in particular to restrict the realized niche (or habitat, see Whittaker 1972) of poorly competitive species to the harshest part of their fundamental or ecophysiological niche (Ellenberg 1956; Austin and Smith 1989). Following the principle of "competitive exclusion" of Gause (1934), competition was thought to strongly affect species richness in the most benign environments, as demonstrated by a number of gradient analyses (e.g., Whittaker 1956).

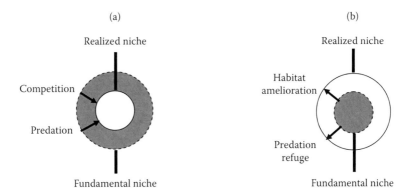

FIGURE 4.1   Inclusion of positive interactions (b) within the niche concept (a). (Adapted from Bruno, Stachowicz, and Bertness. 2003. *Trends in Ecology and Evolution* 18: 119–125.)

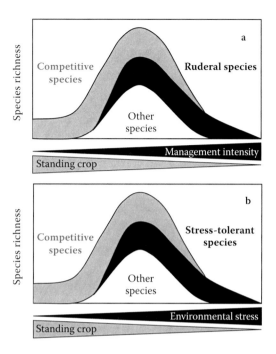

FIGURE 4.2   The humped-back relationship between species richness and standing crop: (a) along a gradient of increasing management intensity and (b) along a gradient of increasing environmental stress. (Adapted from Grime. 1973. *Nature* 242: 344–347.)

Grime (1973) was the first to propose a conceptual model shaping variation in both species richness and biotic interactions along environmental gradients for herbaceous communities. He initially distinguished two main ecological gradients: a gradient of increasing disturbance ("increasing intensity of management," Figure 4.2a) for fertile environments and a gradient of increasing environmental stress and decreasing productivity (Figure 4.2b). Using data from British communities, he obtained a humped-back or unimodal relationship between species diversity and environmental conditions along both gradients, with the highest species richness occurring at intermediate position along the gradients. The decrease in species richness occurring from sites with intermediate levels of stress or disturbance to very stressed or disturbed sites was proposed to be driven only by the species' physiological tolerances to either environmental stress or disturbance but not by biotic processes (right part of the gradients, Figures 4.2a and 4.2b). In sharp contrast, competitive exclusion was thought to regulate diversity in benign environmental conditions (left part of the gradients, Figures 4.2a and 4.2b) due to the niche-shrinking process affecting ruderals and stress-tolerant species (Grime 1974), whereas the effect of the abiotic environment was thought to be minimal.

A number of authors have gathered both humped-back relationships of Grime (1973) within a single graphical model diversely known as the "intermediate disturbance" or "compensatory mortality" hypothesis (Menge and Sutherland 1976;

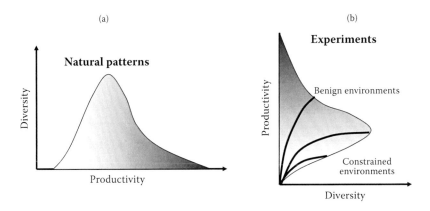

FIGURE 4.3   Hypothesized relationships between (a) diversity–productivity patterns driven by environmental conditions across sites (natural patterns) and (b) the local effect of diversity on productivity (experiments). (Adapted from Loreau et al. 2001. *Science* 294: 804–808.)

Connell 1978; Huston 1979). In this synthetic model, physical disturbance, preda-tion, and physical stress are thought to similarly enhance diversity at intermediate position along a single gradient of mortality. In most textbooks and recent papers on the effect of diversity on community productivity, this relationship is called the uni-modal or humped-back diversity–productivity relationship (e.g., Loreau et al. 2001; see Figure 4.3a). However, in a meta-analysis on the relationship between species richness and productivity, Mittelbach et al. (2001) have stressed that humped-back-shaped curves were especially common (65%) in studies of plant diversity that used plant biomass as a measure of productivity. It should be noted that the humped-back shape proposed by Grime (1973) for British communities was elaborated from results of studies that have also used plant biomass as a surrogate of productivity (Al-Mufti et al. 1977).

Patterns of species richness are also highly dependent on the scale at which they are measured. Mittelbach et al. (2001) have shown that the humped-back shape observed at local and regional scales commonly disappears at a scale larger than a continent (see also Kikvidze et al. 2005). Furthermore, Huston (1999) argued that species interactions are likely to play a strong role in determining richness only at the local scale, whereas other mechanisms (including speciation and extinction) are more likely to affect species richness at larger scales.

The effect of seed limitation has in particular been strongly emphasized by a number of authors (e.g., Tilman 1997; Zobel et al. 2000), and an increasing number of modeling studies have shown that, at intermediate regional scale, stochastic fac-tors may produce patterns of species richness that are similar to those recurrently described at local scale (Rajaniemi et al. 2006). For example, Loreau, Mouquet, and Gonzalez (2003) built a model showing that landscape-driven variation in dis-persal rate drives both species diversity and ecosystem productivity. Similar results were found by Matthiessen and Hillebrand (2006) for benthic microalgal metacom-munities. Helm, Hanski, and Pärtel (2006) emphasized the role of past landscape structure and thus the role of past species pools on community richness. Rajaniemi et al. (2006) showed that species pools (and climate variability) shape biodiversity at regional scale for sand dune communities in Israel, although other processes (competition and the abiotic environment) affect diversity at local scale. Similarly, Freestone and Harrisson (2006) showed that regional richness drives local richness in wetlands, although local factors may also operate. Alternatively, Zeiter, Stampfli, and Newbery (2006) showed that seed limitation was important for biodiversity only in intermediate-productivity sites but not in (a) stressed sites due to the effect of the abiotic conditions (see also Wilsey and Polley 2003) or (b) very productive ones because of the dominant role of competition. Foster and Dickson (2004) also demonstrated that biodiversity and productivity were regulated in grasslands in Kansas by both the biodiversity at the level of the propagule pool and fluctuations of resources, i.e., the abiotic environment, with strong interactive effects.

Grime (1979) also considered that large-scale differences in species pools might also explain why grasslands from calcareous areas are in general more species-rich than grasslands from siliceous areas, independent of the competitive abilities of both species types. He argued that calcareous areas are more spatially represented in southern latitudes and siliceous areas in northern latitudes, where the higher abiotic constraints might have decreased the latter species pool relative to the former during Quaternary species migrations.

However, Michalet et al. (2002) have shown that at local scale the lower species richness of subalpine communities from acidic soils (as compared with communities from calcareous soils) may be explained by differences in species strategies and competitive exclusion processes. Between 2000 and 2100 m of elevation in the French Alps, they analyzed differences in species composition between communities of both soil types using ordination techniques. They found a humped-back relationship between species richness and relevés scores along the first axis of a CA analysis (Figure 4.4); this axis was strongly and significantly correlated to an index of "aboveground space occupancy" (I = vegetation cover × vegetation height; $r = 0.58$, $p < .001$). The highest species richness was found at intermediate position along the gradient on deep and moderately acidic soils localized on calcareous rocks. From this intermediate position, species richness (a) strongly decreased with increasing CA axis 1 relevés scores and aboveground space occupancy on acidic soils localized on siliceous rocks and (b) moderately decreased with decreasing CA axis 1 relevés scores and aboveground space occupancy on dry calcareous soils. Additionally, Choler, Michalet, and Callaway (2001) experimentally measured competitive responses for five species in both a species-poor acidic community and a species-rich calcareous community. They found strong negative interactions for all five species from the acidic community and only one weakly significant competitive response in the calcareous community (Figure 4.5). Michalet et al. (2002) concluded that the lower species richness of acidic soils could very likely be explained by the competitive exclusion of stress-tolerant species from calcareous soils by the competitive species from acidic soils. This increase in community biomass and competitive abilities of species along the gradient was due to the higher water availability of siliceous soils as compared to calcareous soils. This argument was supported by soil-water measurements.

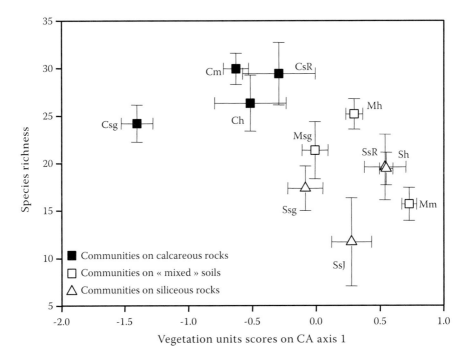

FIGURE 4.4 The humped-back relationship between species richness and the CA axis 1 scores (related to an increasing gradient of aboveground space occupancy) of subalpine communities from the French Alps. (Adapted from Michalet et al. 2002. *Arctic Antarctic & Alpine Research* 34: 102–113.)

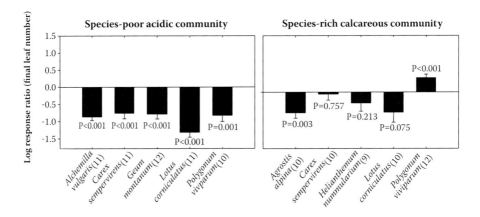

FIGURE 4.5 Interactive responses (positive values indicate facilitation) of species from a species-poor acidic community (Mm in Figure 4.4) and a species-rich calcareous community (Csg in Figure 4.4). (Adapted from Choler, Michalet, and Callaway. 2001. *Ecology* 82: 3295–3308.)

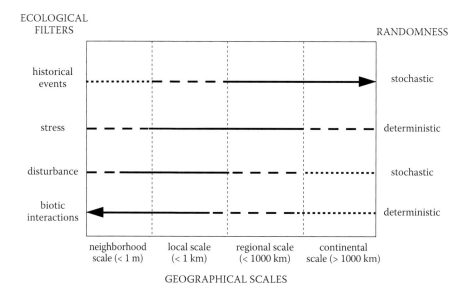

FIGURE 4.6 Changes in the relative importance of the main ecological filters of plant communities with increasing geographical scales.

However, high competition is not always related to decreasing species richness, as demonstrated by Lamb and Cahill (2008) in grassland communities from central Alberta. In plant communities with little shoot competition, root competition and community structure should be unlinked because size-symmetric root competition cannot lead to competitive exclusion by itself. In contrast, in more productive communities, interactions between root and shoot competition may indirectly affect species richness by altering the overall asymmetry of competition (Lamb, Kembel, and Cahill 2009).

To conclude (see Figure 4.6), at local scale aboveground competition is widely recognized as a main driver of species richness together with environmental stress and disturbance, and these effects have even certainly been underestimated by Grime (1979), who was the first to propose the humped-back diversity–biomass model. However, at continental scale, patterns of community richness are primarily driven by chance biogeographical events (e.g., historical events) (Lortie et al. 2004). Recent studies (e.g., Rajaniemi et al. 2006) have emphasized that, at intermediate regional scale, both niche-driven (i.e., deterministic factors in Figure 4.6) and stochastic factors strongly interact to produce patterns of species richness that may, however, be similar to those recurrently described at local scale (i.e., the humped-back diversity-biomass relationship).

4.2.2 FACILITATION AND BIODIVERSITY

Because positive interactions alleviate stress or physical disturbance (i.e., environmental severity), the realized niche of a species (Hutchinson 1959) can be expanded by facilitation (Hacker and Gaines 1997; Choler, Michalet, and Callaway 2001; Bruno, Stachowicz, and Bertness 2003; see Figure 4.1b), which in turn might increase species richness in stressful and/or physically disturbed environments, as shown in salt marshes (Hacker and Gaines 1997), in alpine grasslands (Kikvidze and Nakhutsrishvili 1998; Cavieres et al. 2002, 2006), or arid and semiarid environments (Anthelme, Michalet, and Saadou 2007; Valiente-Banuet et al. 2006). Numerous other examples of this effect of facilitation for biodiversity are provided in Chapter 3, this volume.

Despite clearly relevant recent advances in understanding the role of facilitation for biodiversity, there have been only two previous attempts (Hacker and Gaines 1997; Michalet et al. 2006) to include facilitation within the humped-back model of the diversity–biomass relationship of Grime (1973). Hacker and Gaines (1997) proposed that positive interactions increase species diversity by directly facilitating species that might not normally survive under very high physical disturbance, stress, or predation. They proposed a conceptual scheme, consistent with Bertness and Callaway (1994), in which the positive effects on biodiversity are increasing from intermediate to very high environmental severity (Figure 4.7). As a result, their inclusion of facilitation within the humped-back model did not fundamentally explain the decrease in species richness occurring in the right part of the gradient. Conversely to the effect of competition proposed by Grime (1973) for the left side of the model, facilitation is only assumed to skew the curve but not to drive its direction. This is not surprising because, at the time when Hacker and Gaines (1997) proposed their

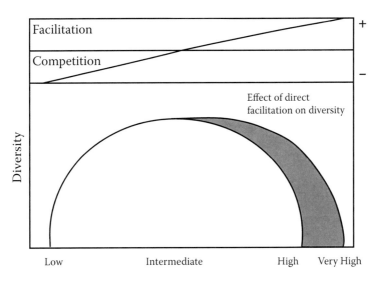

FIGURE 4.7 Inclusion of direct facilitation within the humped-back model (lower panel; adapted from Hacker and Gaines [1997]) and the corresponding changes in biotic interactions following Bertness and Callaway (1994).

model, facilitation was assumed to increase infinitely with environmental severity (Bertness and Callaway 1994) and thus was negatively correlated with community richness in the humped-back model (Figure 4.7).

Michalet et al. (2006) argued that this shortcoming occurs because early facilitation models did not consider extremely severe environments (see Bertness and Callaway 1994; Brooker and Callaghan 1998). However, recent experimental studies indicate that the role of facilitation may actually decrease in exceptionally severe environments (Kitzberger, Steinaker, and Veblen 2000; Anthelme, Michalet, and Saadou 2007; Bruno, Stachowicz, and Bertness 2003). For example, in their intercontinental study of biotic interactions along altitudinal gradients in alpine and arctic communities, Callaway et al. (2002) found also that facilitation was less intense in the arctic sites than in the temperate sites, although the former were more stressful than the latter. This suggests that the importance or intensity of facilitation may actually decrease with increasing stress or physical disturbance in the most severe conditions.

In light of this new evidence, Michalet et al. (2006) proposed another inclusion of facilitation within the humped-back diversity–biomass relationship of Grime (1973). In this model (Figure 4.8), the role of facilitation peaks in conditions of intermediate severity, particularly at the point of highest species richness along the environmental gradient. When stress and/or physical disturbance increases from this intermediate

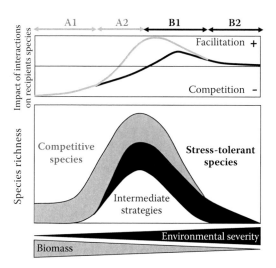

FIGURE 4.8 Inclusion of facilitation into Grime's humped-back model of the relationship between species richness and both biomass and environmental severity. Environmental severity includes both stress and physical disturbance, consistently with Bertness and Callaway (1994). Lower panel: species richness within the community of three strategy types, namely competitive species (gray), stress-tolerant species (black), and plants with intermediate strategies (white), e.g., C-S, sensu Grime (1974). (Adapted from Grime. 1973. *Nature* 242: 344–347.) Upper panel: the average type of net interactions (the sum of positive and negative interactions between neighbors) being received by competitive species (gray curve) and stress-tolerant species (black curve). (Adapted from Michalet et al. 2006. *Ecology Letters* 9: 767–773.)

point (part B1 of the gradient, Figure 4.8), the positive effects of the benefactors decrease, probably because they are less successful at ameliorating abiotic conditions and promoting the survival of beneficiary species in these very severe environments (Kitzberger, Steinaker, and Veblen 2000). In the most severe environmental conditions (part B2 of the gradient, Figure 4.8), biotic interactions become unimportant relative to the effect of the environment. Diversity also decreases with decreasing environmental severity from the intermediate position along the gradient (part A2, Figure 4.8) due to competitive exclusion, as initially proposed by Grime (1973). This model is also in agreement with the results of recent modeling studies that showed that increasing stress and/or physical disturbance may induce a collapse of facilitation and catastrophic shifts in diversity in patchy arid communities (Rietkerk et al. 2004; Kéfi et al. 2007).

Michalet et al. (2006) proposed that the contrasting effects of competition and facilitation for species with different life-history traits combine to produce the humped-back relationship (two different curves in Figure 4.8). In particular, the upper panel of the Figure 4.8 shows how interactions change for different strategy groups along an environmental-severity gradient. At certain parts of the gradient, competitive species can be facilitated, while stress-tolerant species will suffer competition (Figure 4.8, parts A2 and B1). This co-occurrence within a given community of competitive and facilitative responses for stress-tolerant and competitive species, respectively, has been observed by several authors in a number of intermediate- to high-severity environments (Choler, Michalet, and Callaway 2001; Gómez-Aparicio et al. 2004; Liancourt, Callaway, and Michalet 2005). For example, in the calcareous-soil grasslands of Europe, Liancourt et al. (2005) quantified biotic interactions in two conditions of water availability for three dominant species of contrasted strategies. In an unwatered mesophilic and species-rich *Bromus erectus* community, the two competitive species, *Arrhenatherum elatius* and *Brachypodium rupestre*, were highly facilitated for survival, whereas the stress-tolerant species *Bromus erectus* was only slightly negatively affected by neighbors. Conversely, in watered conditions simulating the species-poor *Brachypodium rupestre* community, there were no significant interactions for the survival of the two competitive species, whereas the stress-tolerant *Bromus erectus* experienced high competition. This evidence supports Michalet et al.'s model, in which facilitation is the highest for competitive species at the peak of biodiversity within the humped-back model (Figure 4.8).

4.3   BIODIVERSITY AND ECOSYSTEM FUNCTION: THE ROLE OF BIOTIC INTERACTIONS

In light of the fact that the current loss of biodiversity may be detrimental to ecosystem functioning, a number of experimental studies have been conducted in the last decade to address the role of biodiversity for community productivity, stability, and invasibility. Most of this research has concerned herbaceous ecosystems, and the results have shown a positive effect of either the number of species or the number of functional groups on community productivity (Naeem et al. 1994; Tilman and Downing 1994; Tilman, Wedin, and Knops 1996; Hector et al. 1999; Mulder et al. 2002; van Ruijven and Berendse 2005), although others conversely found that changes in productivity were primarily driven by the functional composition of the natural communities or of the experimental assemblages (Wardle et al. 1997; Hooper and Vitousek 1997; Tilman et al. 1997; Bruno et al. 2005). In all of these research studies, biotic interactions were considered the main processes through which diversity or functional composition affects either productivity or stability. Two main interaction mechanisms have been emphasized: niche complementarity (Braun-Blanquet 1932) and facilitation (Clements 1916), with both mechanisms having been proposed by early plant ecologists as the two main processes driving the structure and composition of plant communities.

4.3.1   NICHE COMPLEMENTARITY AND SELECTION EFFECT

Niche complementarity can be defined as a competition reduction process occurring in diverse assemblages because of the high functional dissimilarity of individuals. Diverse assemblages are expected to better use the whole resources of the ecosystem as compared with monocultures, which increases the productivity of the community.

Braun-Blanquet (1932) was the first to consider that this process was driving the coexistence of a high number of species in most natural plant communities. Early diversity–productivity experiments have emphasized this process to explain their observed positive effect of diversity on productivity (Naeem et al. 1994; Tilman and Downing 1994; Tilman, Wedin, and Knops 1996; Hector et al. 1999). However, others who found a primary role of the functional composition (Hooper and Vitousek 1997; Wardle et al. 1997) have argued that this increase in productivity was in fact due to a selection or sampling effect because of the increasing probability of including a highly productive species when increasing the number of species (Huston 1997). This fueled a substantial controversy among plant ecologists (Huston et al. 2000), but also a higher rigor in diversity–productivity experiments, and several modeling initiatives demonstrated that both effects can be observed and separated (Tilman, Lehman, and Thomson 1997; Loreau and Hector 2001). More recent experiments were able to adequately separate these effects and consistently demonstrated niche-complementarity-driven increases in productivity (van Ruijven and Berendse 2005), although others found that these increases were outweighed by negative selection effects (Hooper and Dukes 2004).

4.3.2   FACILITATION

Two main mechanisms of facilitation may also explain a positive effect of diversity on productivity. The more common observed effect is the improvement of nutrients by legumes, which is a species-specific facilitation (Hector et al. 1999; Tilman et al. 2001; Mulder et al. 2002; Hooper and Dukes 2004). Specific positive effects have also been observed in marine communities (Bruno et al. 2005). However, other experiments have demonstrated that nonspecific facilitation may also drive a positive effect of biodiversity on community productivity (e.g., Mulder, Uliassi, and Doak 2001; and see other examples in Chapter 1, this volume).

4.4 RECONCILING DIVERSITY EXPERIMENTS AND NATURAL PATTERNS

4.4.1 THE DEBATE

Loreau et al. (2001) argued that the positive relationship between species richness and productivity observed in recent experiments has aroused a high controversy among plant ecologists because this result seemed counter to patterns often observed in nature, where the most productive ecosystems are typically characterized by low species diversity, as suggested by the humped-back curves of the literature (Grime 1973; Huston 1979). Loreau et al. (2001) proposed to reconcile natural patterns with experiments by considering that spatial patterns reveal correlations between diversity and productivity driven by environmental factors (Figure 4.3a), whereas small-scale experiments reveal the effect of species properties and diversity on productivity that are detected after the effects of other environmental factors have been removed (Figure 4.3b).

We suggest that there is a substantial inconsistency in this proposition of Loreau et al. (2001) for two reasons. First, at the neighborhood scale (i.e., in diversity experiments), they consider that species diversity directly drives the increase in productivity mainly through reduced competition (niche complementarity) and that this effect operates at all levels of natural productivity along environmental gradients (see Figure 4.3b). This is inconsistent with Grime (1974), who proposed that competition is only important and intense in productive communities. In other words, if there is no competition in harsh unproductive environments, there is no reason that competition should decrease in intensity when increasing species richness in experiments conducted in these environmental conditions. Second, they consider that the large-scale diversity/productivity relationship described in the humped-back model is not causal, but that both diversity and productivity are driven by the environment at this scale. This is also inconsistent with Grime (1973), because in the humped-back model the decrease in species richness in the most productive environments is due to increased competition, which means that the relationship between productivity and diversity is causal in this model through changes in the importance and intensity of competition.

This proposition of Loreau et al. (2001) of two main independent processes operating in plant communities at different scales has already been visited by Levine (2000) and Naeem et al. (2000) to explain the inconsistency existing between the well-known large-scale positive relationship between native species richness and invasive species richness (Planty-Tabacchi et al. 1996), and the negative effect of diversity for invasibility observed in experiments. In this latter example, the authors have stressed that the driving large-scale effect of the environment for both native and invasive species richness was a stochastic effect related to species dispersal abilities and the geography of the environment. However, in the case of the diversity/productivity debate, Loreau et al. (2001) argued that all environmental factors (including deterministic environmental factors such as soil and climate) are driving this noncausal humped-back relationship between species richness and productivity (see Figure 4.3b). In the theoretical proposition of Naeem et al. (2000) and Levine (2000), there is no inconsistency because stochastic environmental factors driving species richness at continental and regional scales have no influence on competition mechanisms operating at the neighborhood scale. In contrast, deterministic environmental factors operating at local scales are known to strongly influence biotic interactions in the humped-back relationship of Grime (1973).

In this last section we will propose an alternative reconciliation of diversity experiments with natural patterns, first for constrained unproductive environments, and second for benign productive environments.

4.4.2 CONSTRAINED ENVIRONMENTS AND FACILITATION

The inclusion of facilitation within Grime's humped-back model of species richness proposed by Michalet et al. (2006) may be particularly helpful for explaining some of the discrepancies observed between natural patterns of species richness and diversity experiments. If facilitation, biodiversity, and both biomass and productivity are all positively correlated along a part of the gradient of environmental severity (parts B1 and B2 of the gradient of Figure 4.8), we can then explain why some experimental studies conducted in stressful conditions found that diversity enhanced productivity.

Specifically, diversity is very likely to increase productivity when facilitation is present, but not when it is absent from an assemblage of species, which itself depends on the position of the community along the environmental gradient. For example, Mulder, Uliassi, and Doak (2001) demonstrated that the productivity of experimental bryophyte assemblages significantly increased with diversity in simulated water-stressed conditions, but was not affected by diversity in controlled wet conditions. Furthermore, they showed that species survival increased with diversity in stressed conditions, which means that the increase in productivity with diversity was due to facilitation and not to either niche-complementarity processes or a sampling effect (see Chapter 1, this volume, for more detailed arguments).

4.4.3 PRODUCTIVE ENVIRONMENTS AND NICHE COMPLEMENTARITY

The great controversy about the results of diversity experiments arose among plant ecologists mainly because it was largely assumed in the literature that diversity was negatively related to productivity in natural benign environmental conditions (Loreau et al. 2001). We consider that the widely accepted opinion of the literature—that species richness is overall negatively related to productivity in productive environments—is due to the use of biomass as a surrogate of productivity, as previously suggested by Mittlebach et al. (2001). In the original humped-back model of Grime (1973), diversity was shown to decrease with increasing standing crop in undisturbed fertile conditions, the result of increasing competition with decreasing disturbance (gradient of management intensity in Figure 4.2a). However, productivity was not measured along this gradient in the original study of Grime (1973; see also Al-Mufti et al. 1977), and Mittlebach et al. (2001) argued that the humped-back relationship generally disappeared when authors used productivity measurements instead of biomass (i.e., standing crop). We will present original data of Touzard, Clément, and Lavorel (2002) on changes in species diversity along successional gradients in eutrophic alluvial grasslands and old fields from French Brittany to analyze the relationship existing between biomass, productivity, and diversity along management-intensity gradients in benign environmental conditions.

Touzard, Clément, and Lavorel (2002) analyzed variation in species richness, diversity, and evenness for 150 plots of mowed and abandoned grasslands (5 to 20 years of abandonment) located in homogeneous fertile conditions in an alluvial eutrophic landscape of French Brittany. Measurements of the total aboveground biomass (live + litter) were randomly sampled in the period of highest standing crop (July) in 34 of the 150 plots (18 grasslands and 16 old fields). Live biomass could be assimilated to productivity in this system because no living biomass is present during the cold season (Blaise Touzard, personal observation). Productivity of grasslands was significantly higher than that of old fields, but standing crop was significantly greater for old fields than for grasslands (Figure 4.9). Although live biomass was 30% lower in old fields, total biomass was 80% higher than in mowed grasslands because litter mass in old fields was 50% higher than live biomass, while there was no litter in mowed grasslands (Figure 4.9). Overall, species richness and species diversity were both 80% higher and species evenness 35% higher in grasslands than in old fields (Figure 4.10). Species richness was highly significantly negatively related to standing crop (Figure 4.11), consistent with the humped-back model of Grime (1973), whereas species richness nonsignificantly increased with increasing productivity (data not shown). The decrease in species richness with increasing standing crop in old fields was mainly driven by the litter accumulation, as already shown by Foster (1999) in North American old fields. Although live biomass was lower in old fields than in mowed grasslands, competition for light by tall competitors has certainly also

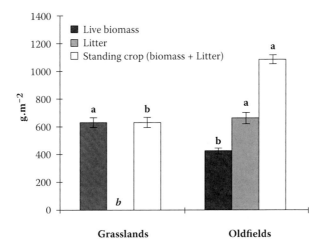

FIGURE 4.9 Live biomass (g×m^{-2}), litter (g×m^{-2}), and standing crop (g×m^{-2}) in old fields and grasslands. Error bars indicate the standard error of the mean. The different styles of letters (lightface, italic, and boldface) indicate significant differences between the means at $p = .05$. The lightface letters indicate the result of the mean comparisons between grasslands and old fields for live biomass. Italic and boldface letters indicate the results for litter and for standing crop, respectively.

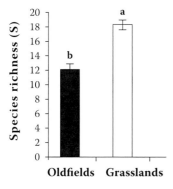
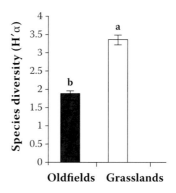
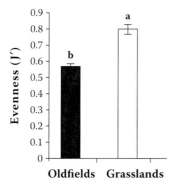

FIGURE 4.10 Species richness S (number of species sampled in a plot), diversity H" (using the Shannon Index; see Shannon and Weaver 1949), and evenness J' (Pielou 1966) in grasslands and old fields. Error bars give the standard error of the mean. Different letters indicate significant differences between the means at $p = .05$.

driven the exclusion of ruderal species in old fields with decreasing disturbance, as proposed by Grime (1973).

The vertical structure of both community types, as assessed with point contact measurements (Figure 4.12), showed that the old fields were dominated by tall species, with a low representation of other species below the canopy of the dominants, whereas in the grasslands the vertical distribution was more even, with a higher representation of short species owing to the ruderal strategy type of Grime (1974). The higher overall productivity of mowed grasslands, as compared with old fields, may be explained by several processes, and in particular the increase in nitrogen content of species with disturbance, as shown by Ryser and Urbas (2000) and Craine et al. (2002). However, we would like to stress the influence of niche-complementarity processes, which have been rarely involved in the literature, likely because relationships between diversity and productivity along natural gradients have been assessed before the emergence of diversity-productivity experiments.

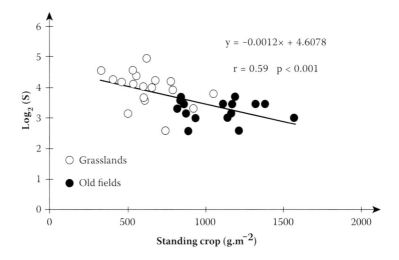

FIGURE 4.11   Linear regression between species richness ($\log_2 S$) and standing crop (biomass + litter) for grasslands and old fields.

In the Ecotron pioneer study of Naeem et al. (1994), the authors argued that the increase in the vertical evenness of the community with increasing diversity was certainly the driving factor for increasing productivity through an increase in resource uptake by niche-complementarity processes (reduced competition for light). Our point-measurements data showed that vertical evenness (and horizontal evenness as

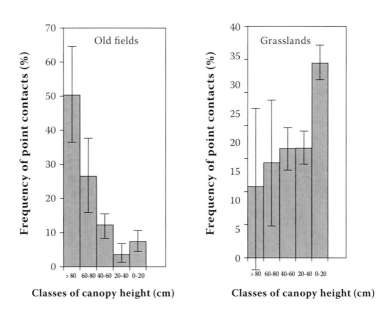

FIGURE 4.12   Height classes distributions in old fields and grasslands.

well) also increased from species-poor old fields to species-rich mowed grasslands in a natural landscape. Measurements of nutrient uptake of the different species involved in those natural assemblages should be done to test the hypothesis that increases in productivity along natural gradients of management intensity in benign environmental conditions are in part driven by a positive niche-complementarity effect of diversity, as recurrently described in diversity experiments. This may reconcile natural patterns with diversity experiments in benign environments.

REFERENCES

Al-Mufti, M. M., C. L. Sydes, S. B. Furness, J. P. Grime, and S. R. Band. 1977. A quantitative analysis of shoot phenology and dominance in herbaceous vegetation. *Journal of Ecology* 65: 759–791.

Anthelme, F., R. Michalet, and M. Saadou. 2007. Positive associations involving the tussock grass *Panicum turgidum* Forssk. in the Aïr-Ténéré Natural Reserve, Niger. *Journal of Arid Environment* 68: 348–362.

Austin, M. P., and T. M. Smith. 1989. A new model for the continuum concept. *Vegetatio* 83: 35–47.

Bertness, M. D., and R. M. Callaway. 1994. Positive interactions in communities. *Trends in Ecology and Evolution* 9: 191–193.

Braun-Blanquet, J. 1932. *Plant sociology, the study of plant communities*. Trans. G. D. Füller and H. S. Connard. New York: McGraw-Hill.

Brooker, R. W., and T. V. Callaghan. 1998. The balance between positive and negative interactions and its relationship to environmental gradient: A model. *Oikos* 81: 196–207.

Bruno, J. F., K. E. Boyer, J. E. Duffy, S. C. Lee, and J. S. Kertesz. 2005. Effects of macroalgal species identity and richness on primary production in benthic marine communities. *Ecology Letters* 8: 1165–1174.

Bruno, J. F., J. J. Stachowicz, and M. D. Bertness. 2003. Inclusion of facilitation into ecological theory. *Trends in Ecology and Evolution* 18: 119–125.

Callaway, R. M., R. W. Brooker, P. Choler, Z. Kikvidze, C. J. Lortie, R. Michalet, L. Paolini, et al. 2002. Positive interactions among alpine plants increase with stress. *Nature* 417: 844–848.

Cavieres, L., M. T. K. Arroyo, A. Peñaloza, M. Molina-Montenegro, and C. Torres. 2002. Nurse effect of *Bolax gummifera* cushion plants in the alpine vegetation of the Chilean Patagonian Andes. *Journal of Vegetation Science* 13: 547–554.

Cavieres, L. A., E. I. Badano, A. Sierra-Almeida, S. Gómez-González, and M. Molina-Montenegro. 2006. Positive interactions between alpine plant species and the nurse cushion plant *Laretia acaulis* do not increase with elevation in the Andes of Central Chile. *New Phytologist* 169: 59–69.

Chapin, F. S. I., E. S. Zavaleta, V. T. Eviner, R. L. Naylor, P. M. Vitousek, H. L. Reynolds, D. U. Hooper, et al. 2000. Consequences of changing biodiversity. *Nature* 405: 234–242.

Choler P., R. Michalet, and R. M. Callaway. 2001. Facilitation and competition on gradients in alpine plant communities. *Ecology* 82: 3295–3308.

Clements, F. E. 1916. *Plant succession*. Washington, DC: Carnegie Institution.

Connell, J. H. 1978. Diversity in tropical rain forests and coral reefs. *Science* 199: 1302–1310.

Craine, J. M., D. Tilman, D. Wedin, P. Reich, M. Tjoelker, and J. Knops. 2002. Functional traits, productivity and effects on nitrogen cycling of 33 grassland species. *Functional Ecology* 16: 563–574.

Diaz, S., S. Lavorel, F. de Bello, F. Quétier, K. Grigulis, and T. M. Robson. 2007. Incorporating plant functional diversity effects in ecosystem service assessments. *Proceedings National Academy of Sciences USA* 104: 20684–20689.

Ellenberg, H. 1956. Aufgaben und Methoden der Vegetationkunde. In *Einführung in die Phytologie*, vol. 4, pt. 1, ed. H. Walter. Stuttgart: Ulmer.

Foster, B. L. 1999. Establishment, competition and the distribution of native grasses among Michigan old-fields. *Journal of Ecology* 87: 476–489.

Foster, B. L., and T. L. Dickson. 2004. Grassland diversity and productivity: The interplay of resource availability and propagule pools. *Ecology* 85: 1541–1547.

Freestone, A. L., and S. Harrisson. 2006. Regional enrichment of local assemblages is robust to variation in local productivity, abiotic gradients, and heterogeneity. *Ecology Letters* 9: 95–102.

Gause, G. F. 1934. *The struggle for existence*. New York: Haffner.

Gomez-Aparicio, L., R. Zamora, J. M. Gomez, J. A. Hodar, J. Castro, and E. Baraza. 2004. Applying plant positive interactions to reforestation of Mediterranean mountains: A meta-analysis of the use of shrubs as nurse plants. *Ecological Applications* 14: 1128–1138.

Grime, J. P. 1973. Competitive exclusion in herbaceous vegetation. *Nature* 242: 344–347.

Grime, J. P. 1974. Vegetation classification by reference to strategies. *Nature* 250: 26–31.

Grime, J. P. 1979. *Plant strategies and vegetation processes*. Chichester, England: Wiley & Sons.

Hacker, S. D., and S. D. Gaines. 1997. Some implications of direct positive interactions for community species diversity. *Ecology* 78: 1990–2003.

Hector, A., B. Schmid, C. Beierkuhnlein, M. C. Caldeira, M. Diemer, P. G. Dimitrakopoulos, J. A. Finn, et al. 1999. Plant diversity and productivity experiments in European grasslands. *Science* 286: 1123–1127.

Helm, A., I. Hanski, and M. Pärtel. 2006. Slow response of plant species richness to habitat loss and fragmentation. *Ecology Letters* 9: 72–77.

Hooper, D. U., F. S. Chapin III, J. J. Ewel, A. Hector, P. Inchausti, S. Lavorel, J. H. Lawton, et al. 2005. Effects of biodiversity on ecosystem functioning: A consensus of current knowledge. *Ecological Monographs* 75: 3–35.

Hooper, D. U., and J. S. Dukes. 2004. Overyielding among functional groups in a long-term experiment. *Ecology Letters* 7: 95–105.

Hooper, D. U., and P. M. Vitousek. 1997. The effect of plant composition and diversity on ecosystem processes. *Science* 277: 1302–1305.

Huston, M. A. 1979. A general hypothesis of species diversity. *American Naturalist* 113: 81–101.

Huston, M. A. 1997. Hidden treatments in ecological experiments: Re-evaluating the ecosystem function of biodiversity. *Oecologia* 110: 449–460.

Huston, M. A. 1999. Local processes and regional processes: Appropriate scales for understanding variation in the diversity of plants and animals. *Oikos* 86: 393–401.

Huston, M. A., L. W. Aarssen, M. P. Austin, B. S. Cade, J. D. Fridley, E. Garnier, J. P. Grime, et al. 2000. No consistent effect of plant diversity on productivity. *Science* 289: 1255.

Hutchinson, G. E. 1959. Homage to Santa Rosalia, or why are there so many kinds of animals? *American Naturalist* 93: 145–159.

Kéfi, S., M. Rietkerk, C. L. Alados, Y. Pueyo, V. P. Papanastasis, A. ElAich, and P. C. de Ruiter. 2007. Spatial vegetation patterns and imminent desertification in Mediterranean arid ecosystems. *Nature* 449: 213–217.

Kikvidze, Z., and G. Nakhutsrishvili. 1998. Facilitation in subnival vegetation patches. *Journal of Vegetation Science* 9: 261–264.

Kikvidze, Z., F. I. Pugnaire, R. W. Brooker, P. Choler, C. J. Lortie, R. Michalet, and R. M. Callaway. 2005. Linking patterns and processes in alpine plant communities: A global study. *Ecology* 86: 1395–1400.

Kitzberger, T., D. F. Steinaker, and T. T. Veblen. 2000. Effects of climatic variability on facilitation of tree establishment in northern Patagonia. *Ecology* 81: 1914–1924.

Lamb, E. G., and J. F. Cahill. 2008. When competition does not matter: Grassland diversity and community composition. *American Naturalist* 171: 777–787.

Lamb, E. G., S. W. Kembel, and J. F. Cahill. 2009. Shoot, but not root, competition reduces community diversity in experimental mesocosms. *Journal of Ecology* 97: 155–163.

Levine, J. M. 2000. Species diversity and biological invasions: Relating local process to community pattern. *Science* 288: 852–854.

Liancourt, P., R. M. Callaway, and R. Michalet. 2005. Stress tolerance and competitive-response ability determine the outcome of biotic interactions. *Ecology* 86: 1611–1618.

Loreau, M., and A. Hector. 2001. Partitioning selection and complementarity in biodiversity experiments. *Nature* 412: 72–76.

Loreau, M., N. Mouquet, and A. Gonzalez. 2003. Biodiversity as spatial insurance in heterogeneous landscapes. *Proceedings of the National Academy of Sciences USA* 100: 12765–12770.

Loreau, M., S. Naeem, P. Inchausti, J. Bengtsson, J. P. Grime, A. Hector, D. U. Hooper, et al. 2001. Biodiversity and ecosystem functioning: Current knowledge and future challenges. *Science* 294: 804–808.

Lortie, C. J., R. W. Brooker, P. Choler, Z. Kikvidze, R. Michalet, F. I. Pugnaire, and R. M. Callaway. 2004. Rethinking plant community theory. *Oikos* 107: 433–438.

Matthiessen, B., and H. Hillebrand. 2006. Dispersal frequency affects local biomass production by controlling local diversity. *Ecology Letters* 9: 652–662.

Menge, B. A., and J. P. Sutherland. 1976. Species diversity gradients: Synthesis of the roles of predation competition and temporal heterogeneity. *American Naturalist* 110: 351–369.

Michalet, R., R. W. Brooker, L. A. Cavieres, Z. Kikvidze, C. J. Lortie, F. I. Pugnaire, A. Valiente-Banuet, and R. M. Callaway. 2006. Do biotic interactions shape both sides of the humped-back model of species richness in plant communities? *Ecology Letters* 9: 767–773.

Michalet, R., C. Gandoy, D. Joud, J. P. Pages, and P. Choler. 2002. Plant community composition and biomass on calcareous and siliceous substrates in the northern French Alps: Comparative effects of soil chemistry and water status. *Arctic Antarctic & Alpine Research* 34: 102–113.

Mittelbach, G. G., C. F. Steiner, S. M. Scheiner, K. L. Gross, H. L. Reynolds, R. B. Waide, M. R. Willig, S. I. Dodson, and L. Gough. 2001. What is the observed relationship between species richness and productivity? *Ecology* 82: 2381–2396.

Mulder, C. P. H., A. Jumpponen, P. Högberg, and K. Huss-Danell. 2002. How plant diversity and legumes affect nitrogen dynamics in experimental grassland communities. *Oecologia* 133: 412–421.

Mulder, C. P. H., D. D. Uliassi, and D. F. Doak. 2001. Physical stress and diversity-productivity relationships: The role of positive interactions. *Proceedings of the National Academy of Sciences USA* 98: 6704–6708.

Naeem, S., J. M. H. Knops, D. Tilman, K. M. Howe, T. Kennedy, and S. Gale. 2000. Plant diversity increases resistance to invasion in the absence of covarying extrinsic factors. *Oikos* 91: 97–108.

Naeem, S., L. J. Thompson, S. P. Lawler, J. H. Lawton, and R. M. Woodfin. 1994. Declining biodiversity can alter the performance of ecosystems. *Nature* 368: 734–737.

Pielou, E. C. 1966. The measures of diversity in different types of biological collections. *Journal of Theoretical Biology* 13: 131–144.

Planty-Tabacchi, A. M., E. Tabacchi, R. J. Naiman, C. Deferrari, and H. Decamps. 1996. Invasibility of species rich communities in riparian zones. *Conservation Biology* 10: 598–607.

Rajaniemi, T. K., D. E. Goldberg, R. Turkington, and A. R. Dyer. 2006. Quantitative partitioning of regional and local processes shaping regional diversity patterns. *Ecology Letters* 9: 121–128.

Rietkerk, M., S. C. Dekker, P. C. De Ruiter, and J. Van de Koppel. 2004. Self-organized patchiness and catastrophic shifts in ecosystems. *Science* 305: 1926–1929.

Ryser, P., and P. Urbas. 2000. Ecological significance of life span among Central European grass species. *Oikos* 91: 41–50.

Schröter, D., W. Cramer, R. Leemans, I. C. Prentice, M. B. Araújo, N. W. Arnell, A. Bondeau, et al. 2005. Ecosystem service supply and vulnerability to global change in Europe. *Science* 310: 1333–1337.

Shannon, C. E., and W. Weaver. 1949. *The mathematical theory of communication.* Urbana: University of Illinois Press.

Tilman, D. 1997. Community invasibility, recruitment limitation, and grassland biodiversity. *Ecology* 78: 81–92.

Tilman, D., and J. A. Downing. 1994. Biodiversity and stability in grasslands. *Nature* 367: 363–365.

Tilman, D., J. Knops, D. Wedin, P. B. Reich, M. Ritchie, and E. Siemann. 1997. The influence of functional diversity and composition on ecosystem processes. *Science* 277: 1300–1302.

Tilman, D., C. L. Lehman, and K. T. Thomson. 1997. Plant diversity and ecosystem productivity: Theoretical considerations. *Proceedings of the National Academy of Sciences USA* 94: 1857–1861.

Tilman, D., P. B. Reich, J. Knops, D. Wedin, T. Mielke, and C. Lehman. 2001. Diversity and productivity in a long-term grassland experiment. *Science* 294: 843–845.

Tilman, D., D. Wedin, and J. Knops. 1996. Productivity and sustainability influenced by biodiversity in grassland ecosystems. *Nature* 379: 718–720.

Touzard, B., B. Clément, and S. Lavorel. 2002. Successional patterns in a eutrophic alluvial wetland of Western France. *Wetlands* 22: 111–125.

Valiente-Banuet, A., A. Vital Rumebe, M. Verdu, and R. M. Callaway. 2006. Modern Quaternary plant lineages promote diversity through facilitation of ancient Tertiary lineages. *Proceedings of the National Academy of Sciences USA* 109: 16812–16817.

van Ruijven, J., and F. Berendse. 2005. Diversity-productivity relationships: Initial effects, long-term patterns, and underlying mechanisms. *Proceedings of the National Academy of Sciences USA* 102: 695–700.

Wardle, D. A., O. Zackrisson, G. Hornberg, and C. Gallet. 1997. The influence of island area on ecosystem properties. *Science* 277: 1296–1299.

Whittaker, R. H. 1956. Vegetation of the Great Smoky Mountains. *Ecological Monographs* 26: 1–80.

Whittaker, R. H. 1972. Evolution and measurement of species diversity. *Taxon* 21: 213–251.

Wilsey, B. J., and H. W. Polley. 2003. Effects of seed additions and grazing history on diversity and productivity of subhumid grasslands. *Ecology* 84: 920–931.

Zeiter, M., A. Stampfli, and D. M. Newbery. 2006. Recruitment limitation constrains local species richness and productivity in dry grassland. *Ecology* 87: 942–951.

Zobel, M., M. Otsus, J. Liira, M. Moora, and T. Mols. 2000. Is small-scale species richness limited by seed availability or microsite availability? *Ecology* 81: 3274–3282.

5 Arbuscular Mycorrhizae and Plant–Plant Interactions
Impact of Invisible World on Visible Patterns

Mari Moora and Martin Zobel

CONTENTS

5.1 INTRODUCTION

The word *mycorrhiza* (literally, "fungus root") defines the symbiotic relationship between plant roots and soil fungi. The location of the fungal symbiont in the root and its hyphal connections with the soil enable it to influence the absorption of soil-derived nutrients; in most cases, the fungus obtains organic carbon as a recent photosynthate from the plant. The mycorrhizal condition is typical for most plants under most ecological conditions (Smith and Read 1997).

Arbuscular mycorrhizal (AM) fungi (phylum Glomeromycota) are thought to be the oldest group of organisms living in symbiosis with land plants (Redecker, Morton, and Bruns 2000). They are ubiquitous plant root symbionts that can be considered "keystone mutualists" in terrestrial ecosystems, forming a link between biotic and abiotic components with carbon and nutrient fluxes that pass between plants and fungi in the soil (O'Neill, O'Neill, and Norby 1991). Approximately 200 AM fungal species have been identified (Schussler, Schwarzott, and Walker 2001; Rosendahl 2008), based primarily on spore morphology. AM fungi contribute up to 90% of plant P uptake (van der Heijden et al. 2006) and can contribute to enhanced N acquisition under some conditions as well (Hodge, Campbell, and Fitter 2001). In addition, AM fungi may provide protection against fungal pathogens (Borowicz 2001; Maherali and Klironomos 2007).

AM fungal communities have a great impact on plant community structure and composition (Klironomos 2003; van der Heijden et al. 2006; van der Heijden, Bardgett, and van Straalen 2008; Scheublin, van Logtestijn, and van der Heijden 2007; Rosendahl 2008). This impact is realized through the effect of AM on plant interactions. Because AM fungi and particular fungal taxa are patchily distributed in landscapes and communities, the presence or absence of fungi may influence the succession and restoration of plant communities, as well as the success of alien plant species that have invaded new habitats.

5.2 ARBUSCULAR MYCORRHIZAE AND PLANT–PLANT INTERACTIONS

Earlier studies addressing the role of mycorrhizae on plant–plant interactions have relied on experimental assemblies where one plant species is strongly AM dependent and a second independent (Allen and Allen 1984, 1990). Typically, the presence of AM inoculum confers a greater advantage to the mycorrhizal competitor. These experiments show clearly that mycorrhizal fungi amplify competition among more-or-less mycotrophic (or completely nonmycorrhizal) plants. Consequently, these experiments may help us understand the effect of AM on plant interactions in early-successional ecosystems (coarse-scale effect, sensu Hart, Reader, and Klironomos 2003), where local AM fungal populations are obviously dispersal limited and mycotrophic plant species may be disadvantaged due to the lack of proper symbionts. However, such experiments tell us little about ecosystems in which most plants are mycorrhizal and AM fungal inoculum is available.

When plants coexist in late-successional ecosystems, in which AM fungi are abundant and in contact with the roots of most plants, fine-scale processes such as host specificity, AM multifunctionality, and shared mycelial networks might be important determinants of plant coexistence (Hart, Reader, and Klironomos 2003).

5.2.1 EXPERIMENTAL EVIDENCE OF MYCORRHIZAL INTERFERENCE OF PLANT–PLANT INTERACTIONS

The experimental evidence of mycorrhizal interference of plant–plant interactions is mixed. The number of experiments conducted is still relatively low, and only a few combinations of plant and AM fungal taxa have been addressed.

We analyzed studies comparing competitive response of target species in non-mycorrhizal and mycorrhizal experimental systems (Table 5.1), or in experimental systems with different AM taxon compositions (Table 5.2). We classified the effect of AM as negative (amplifying competition) when AM fungal inoculation increased competitive response, i.e., when competing target plants were smaller with AM than without. The effect of AM was classified as positive (balancing competition) when AM fungal inoculation decreased the competitive response of target plants. When there were no differences between competitive response in mycorrhizal and nonmycorrhizal conditions, the effect was considered neutral (no effect on competition).

We found a total of 17 studies in which 32 interspecific and 16 intraspecific plant–plant interactions were addressed in conditions where AM fungi were present/absent (Table 5.1). We performed a meta-analysis of the recorded responses presented in Table 5.1 using log-linear analysis and a Freeman–Tukey deviation (FTD) test (Legendre and Legendre 1998) with Statistica 6.0 software (StatSoft 2001). Type of plant interaction (intraspecific or interspecific), plant life stage (either plants with similar developmental stages or plants with different developmental stages interacted), and the effect of AM on plant–plant interaction (amplifying, neutral, or balancing) were used as categorical factors (Table 5.3).

The best initial model of log-linear analysis included the type of interaction and life stage as the main effects (Pearson chi-square goodness of fit: Chi sq = 9.00, df = 8, $p = .342$). AM symbiosis influenced the outcome of intra- and inter-specific competition differently (Table 5.3). For intraspecific interaction, there were significantly fewer balancing effects and significantly more amplifying effects on competition than anticipated by the null model ($p < .05$, FTD test). For interspecific interactions, there were significantly fewer amplifying effects and significantly more balancing effects on competition than anticipated by the null model ($p < .05$, FTD test). The influence of AM symbiosis differed significantly when the developmental stage of the interacting plants was considered (Table 5.3). When plants of different life stages interacted, there were significantly fewer amplifying effects on competition than expected by the null model ($p < .05$, FTD test).

We may conclude that the presence of AM fungi may either amplify or not influence intraspecific competition, and tends to balance interspecific competition. The balancing effect was particularly evident when the interspecific interactions between adults and seedlings were considered. When mycorrhizal adults and seedlings are interacting, the outcome of the interaction is dependent possibly both on competition for resources and on the facilitation of seedlings by adults, acting as carbon sources for AM fungi and enhancing inoculation of seedlings via growth of a common mycelial network. The existence of such a dual relationship, which obviously exists for ectomycorrhizae (Dickie et al. 2005), needs to be verified in experiments where the development of AM fungal structures in roots and soil will be tracked over time.

We found only four studies that specifically addressed the effect of different AM fungal taxa and their combinations on plant–plant interactions (Table 5.2). All interactions emerging in experimental systems may be classified as competitive, and facilitation was not evident. The effect of AM on plant–plant interaction depended on the taxon of AM fungus or on the taxon composition of AM fungal communities.

TABLE 5.1

Effect of the Presence of AM Fungi on Plant–Plant Interactions among Naturally Coexisting Plants with Comparable Mycorrhizal Susceptibilities Compared with Nonmycorrhizal

Publication	Target Species	Interspecific Competition		Intraspecific Competition	
		Different Life Stages	Similar Life Stages	Different Life Stages	Similar Life Stages
Fitter (1977)	*Lolium perenne*		−		
	Holcus lanatus		−		
Eissenstat and Newman (1990)	*Plantago lanceolata*			0	
Allsopp and Stock (1992)	*Otholobium hirtum*				−
	Aspalathus linearis				−
Hartnett et al. (1993)	*Andropogon gerardii*		+		−
	Elymus canadensis		−		0
Zobel and Moora (1995)	*Fragaria vesca*		0		
	Centaurea jacea		+		
Moora and Zobel (1996)	*Fragaria vesca*	0			
	Prunella vulgaris	+		−	
West (1996)	*Holcus lanatus*		−		+
	Dactylis glomerata		−		+
Moora and Zobel (1998)	*Hypericum perforatum*			−	
Marler, Zabinski, and Callaway (1999)	*Centaurea maculosa*	+	+		−
	Festuca idahoensis	0	−		0
Ronsheim and Anderson (2001)	*Allium vineale*				+
Callaway et al. (2001)	*Centaurea melitensis*		+		
	Nassella pulchra		−		
Callaway et al. (2003)	*Centaurea melitensis*		+ & 0		
	Nassella pulchra		−		
	Avena barbata		+		
Carey, Marler, and Callaway (2004)	*Centaurea maculosa*	+ & +			
	Festuca idahoensis	0			
	Bouteloua gracilis	0			
Ayres, Gange, and Aplin (2006)	*Plantago lanceolata*				0

TABLE 5.1 (CONTINUED)
Effect of the Presence of AM Fungi on Plant–Plant Interactions among Naturally Coexisting Plants with Comparable Mycorrhizal Susceptibilities Compared with Nonmycorrhizal

Publication	Target Species	Interspecific Competition		Intraspecific Competition	
		Different Life Stages	Similar Life Stages	Different Life Stages	Similar Life Stages
Endlweber and Scheu (2007)	*Lolium perenne*		0		0
	Trifolium repens		0		0
Pietikainen and Kytöviita (2007)	*Gnaphalium norvegicum*			–	

Publication		Target Species						
Callaway et al. (2004): interspecific competition/ similar stage	*Centaurea maculosa*	*Pseudoroegneria spicata*	*Festuca idahoensis*	*Koeleria cristata*	*Gaillardia aristata*	*Linum lewisii*	*Achillea millefolium*	
	–		+	+	–	0	0	

Note: –, the competitive suppression of the target plant increased when AMF were present; +, the competitive suppression of the target plant decreased when AMF were present; 0, no differences between mycorrhizal and nonmycorrhizal treatments.

TABLE 5.2
Influence of Different AM Fungal Taxa or Communities on Plant–Plant Interaction

Publication	Target Species	Experimental Mycorrhizal Fungal Community				
		NM	Basel Bi	BEG 21	BEG 19	Mixture AMF
van der Heijden, Wiemken, and Sanders (2003)	*Prunella vulgaris*	–	+	–	+	–
	Holcus lanatus	+	–	+	–	+

Publication	Target Species	DD-1	DD-2	BEG 21	DD-3	Competition Treatment
Scheublin, van Logtestijn, and van der Heijden (2007)	*Lotus corniculatus*	+	+	+	+	Intraspecific
		+	+	+	+	Interspecific
	Festuca ovina	+	+	+	+	Intraspecific
		–	–	–	–	Interspecific
	Plantago lanceolata	0	0	+	0	Intraspecific
		0	0	0	0	Interspecific

(continued)

TABLE 5.2 (CONTINUED)
Influence of Different AM Fungal Taxa or Communities on Plant–Plant Interaction

Publication	Target Species	Experimental Mycorrhizal Fungal Community				
Landis, Gargas, and Givinish (2005)		*Glomus claroideum*	*Glomus mosseae*	*G. cla.* + *G. mos.*	**Wild AMF Community**	**Competition Treatment**
	Amorpha canescens	–	–	–	0	Intraspecific
		0	0	0	0	Interspecific
	Monarda fistulosa	0	0	0	0	Intraspecific
		0	0	0	0	Interspecific
	Rumex acetosella	0	0	0	0	Intraspecific
		0	+	0	0	Interspecific
	Schizachyrium scoparium	0	0	0	0	Intraspecific
		0	0	0	0	Interspecific
Kytöviita, Vestberg, and Tuomi (2003)		*Glomus claroideum*	*Glomus hoi*	*Glomus* spp.		
	Sibbaldia procumbens	+	+	+		Seedlings
		–	–	–		Neighbor
	Solidago virgaurea	+	0	+		Seedlings
		–	0	–		Neighbor
	Antennaria dioica	+	+	0		Seedlings
		–	–	0		Neighbor
	Campanula rotundifolia	0	0	0		Seedlings
		–	–	–		Neighbor

Note: In van der Heijden, Wiemken, and Sanders (2003), the qualitative estimates (+, increase; –, decrease) indicate the biomass proportion in the species mixture. In Scheublin, van Logtestijn, and van der Heijden (2007) and Landis, Gargas, and Givinish (2005), the qualitative estimates (+, increase; 0, no difference; –, decrease) are made in comparison of aboveground plant biomass in the nonmycorrhizal situation separately for intra- and interspecific plant interaction. In Kytöviita, Vestberg, and Tuomi (2003), the qualitative estimates are made in calculating the difference of fresh total shoot mass from the nonmycorrhizal without neighbor grown seedlings when targets are growing with (neighbor) or without (seedlings) adjacent *Sibbaldia procumbens* large plant.

Although the low number of experiments makes it impossible to draw any general conclusions, the current evidence, together with experiments on the differential effect of natural AM fungal communities on plant growth (Moora et al. 2004; Moora, Öpik, and Zobel 2004; Corkidi et al. 2002; Kiers et al. 2000; Frank et al.

TABLE 5.3

Results of the Meta-Analysis of the Studies Presented in Table 5.1

The Influence of AM Symbiosis on the Plant–Plant Competition	Interaction Type (χ^2 = 4.81, p < .028, df = 1)		Life Stages Involved (χ^2 = 11.05, p < .001, df = 1)	
	Interspecific	Intraspecific	Different Life Stages	Similar Life Stages
Amplifying	less	more	less	ND
No impact	ND	ND	ND	ND
Balancing	more	less	ND	ND

Note: According to the log-linear analysis and Freeman–Tukey deviates (FTD) test, the number of studies belonging to each category was either not different (ND) or it was significantly less or greater than that anticipated by the null model (p < .05, FTD test).

2003), support the idea that varying AM fungal community composition may have an effect on plant coexistence in natural conditions. We suggest that inoculation of experimental systems with natural AM communities needs to be used more widely in future studies because AM taxa that have been brought into culture often have a ruderal life cycle, global distribution, and resistance to soil perturbations (Öpik et al. 2006; van der Heijden, Bardgett, and van Straalen 2008), and experiments with easily cultured AM taxa ignore the ecological significance of uncultured and specialist AM fungi.

5.2.2 THE MECHANISM OF AM EFFECTS

There are two potential mechanisms by which AM fungi can affect plant–plant interactions. First, plant species may differ with respect to their response to AM fungi. For instance, van der Heijden et al. (1998) grew different combinations of three plant species and four AM fungal taxa in greenhouses, and found that the same plant species responded differently, in terms of biomass or phosphorus concentration in tissues, to the presence of specific taxa. Other studies have shown that taxa of AM fungi differ markedly in their improvement of plant nutrition and health (e.g., Munkvold et al. 2004). Moora et al. (2004) showed in a greenhouse experiment that the growth response of two congeneric naturally co-occurring plant species was dependent on the habitat-specific AM fungal communities (Figure 5.1). Consequently, the performance of particular plant species in nature may depend on which plant and fungal taxa meet in situ.

   The second potential mechanism is the connection of plant individuals into a common mycelial network (CMN), which may drain carbon or nutrients from some individuals and support others (Simard and Durall 2004). It is not known, however, under what conditions, if any, such a transfer becomes functionally important, since the quantities of interplant transfer of nutrients are small compared with AM-mediated transfer from the soil to the plant, and the amount of transferred carbohydrates is

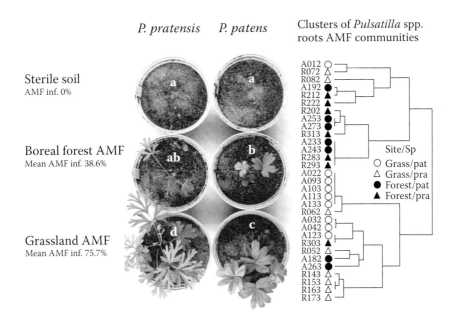

P. pratensis   *P. patens*   Clusters of *Pulsatilla* spp.
roots AMF communities

Sterile soil
AMF inf. 0%

Boreal forest AMF
Mean AMF inf. 38.6%

Grassland AMF
Mean AMF inf. 75.7%

Site/Sp

○ Grass/pat
△ Grass/pra
● Forest/pat
▲ Forest/pra

FIGURE 5.1 (*A color version of this figure follows page 110.*) The effect of boreal pine forest and grassland AMF communities compared with non-AMF containing sterile soil on the performance of *Pulsatilla pratensis* and *P. patens* after 14 weeks growth in a greenhouse. Different lowercase letters above pots indicate a statistically significant difference in total dry mass. Mean AMF infection is the percent of the colonized root length. The results of hier-archical clustering analysis of AM fungal communities in *Pulsatilla* seedling roots are also presented. AM fungi were identified as SSU rDNA sequence groups by Öpik et al. (2003), and each sample (row in the picture) represents the root system of one seedling where the fungal community is described by fungal sequence types present/absent. *Pulsatilla pratensis* (triangles) and *P. patens* (circles) were grown in the presence of forest (filled symbols) and grassland (open symbols) soil inoculum. (Figure is based on Moora et al. [2004]; photo-graph by R. Sen.)

also small (Hart, Reader, and Klironomos 2003). Robinson and Fitter (1999) offered a "mycocentric" view on carbon transfer, claiming that although the carbon is shared among the different plant individuals' mycorrhizae within the same mycelium, the carbon remained within plant roots. Fungal structures within roots thus represent a kind of extended mycelia through which fungi transfer carbon according to their own demands, independently of the needs of autotrophic hosts. Since data on the transfer rates of nutrients and carbon via CMN are still accumulating (Wilson, Hartnett, and Rice 2006), the role of CMN needs further clarification.

In addition to nutrient and carbon transfer, CMN may still act as a specific sup-port system, creating a safe site for the establishment of seedlings that are infected rapidly by common mycelium (van der Heijden 2004). Plant species linked by a CMN may form guilds of mutual aid (Read 1997), and when the majority of plant species in a community are in symbiosis with mycorrhizal fungi, the likelihood of plant establishment increases due to the presence of CMN. At the same time, CMN may inhibit the regeneration of nonmycorrhizal species (Bonis, Grubb, and Coomes 1997). Kytöviita, Vestberg, and Tuomi (2003), however, found that the presence of mycorrhizal adult plants did not improve the growth of seedlings in a greenhouse, although solitary seedlings may benefit from mycorrhizae (Table 5.2). The authors concluded that a CMN may well imply some mutual aid for the connected plants, but competitive interactions among plants overshadow any benefit.

5.3 MULTITROPHIC INTERACTIONS

AM may affect plant–plant interactions indirectly by influencing other organisms in contact with plants. In the following subsections, we review a few possible multitrophic interactions.

5.3.1 OTHER PLANT MUTUALISTS AND MYCORRHIZAE

Many soil-inhabiting bacteria—so-called plant-growth-promoting rhizobacteria (PGPR, including N_2-fixing bacteria)—stimulate plant growth through direct or indirect interactions with plant roots (Artursson, Finlay, and Jansson 2006). These bacteria may support plant growth through the synergistic effects with AM, since in conditions when both nitrogen and phosphorus limit plant growth, AM fungi can improve P uptake and result in more available energy for N fixation (Fitter and Garbaye 1994). This synergism may result in more efficient nutrient acquisition, inhibition of pathogens, and enhancement of root branching (Barea, Azcon-Aguilar, and Azcon 1997; Barea, Azcon, and Azcon-Aguilar 2002).

Interactions among plants and pollinators may be influenced by belowground mutualists as well. For instance, AM fungi can alter the reproductive traits of flowering plants (Koide and Dickie 2002) and may thereby also influence plant interactions with pollinators. Indeed, there is evidence that mycorrhizal plants were more often visited by pollinators than nonmycorrhizal plants (Wolfe, Husband, and Klironomos 2005; Gange and Smith 2005). In a recent field experiment, Cahill et al. (2008) demonstrated the shift in floral visitor community and reduced floral visits per flowering stem across the 23 flowering species found in the community after three years of AM fungal suppression. However, these effects were due to changes in floral-visitor behavior due to altered patch-level floral display, rather than through direct effects of AMF (AM fungi) suppression on floral morphology.

5.3.2 PLANT PREDATORS AND MYCORRHIZAE

Mutual relationships among plant predators (herbivores, fungal and microbial pathogens, parasitic plants) and AM fungi may affect plant–plant relationships. For instance, plant growth and foliar nutrient content are increased by particular AM fungi (Moora et al. 2004; van der Heijden 2004), and mycorrhizal plants can be more attractive to herbivores than nonmycorrhizal plants (Gange and West 1994; Goverde et al. 2000). When mycorrhizal plants are preferred by predators, nonmycorrhizal neighbors may benefit by being less attractive and gain dominance. It is also possible, however, that the presence of AM fungi reduces the attack of predators. For instance, the presence of AM fungi in roots may reduce the severity of root damage by soilborne pathogens (Kjøller and Rosendahl 1996; Jaizme-Vega et al. 1997; Colditz et al. 2005). This effect may also improve the nutritional status of plants. For example, better nutritional status of plants may result in increased production of secondary metabolites (Colditz et al. 2005). Some predators consume AM fungal spores and may thus also have a positive effect on plants, since they act as spore dispersers (Janos, Sahley, and Emmons 1995; Mangan and Adler 2002).

All previous examples describe theoretical possibilities of how the effect of AM fungi on plant–plant interactions may be influenced by multitrophic relationships. We are aware of only one study (Pietikainen and Kytoviita 2007) in which the simultaneous effect of simulated herbivory and AM fungi on plant–plant interactions was estimated. In that study, the defoliation of large neighbors increased the beneficial effect of AM on seedlings.

5.4   LARGE-SCALE CHANGES IN PLANT COMMUNITY STRUCTURE

5.4.1   Mycorrhizae and Succession

Succession is often viewed as a competitive displacement of competitively weaker pioneer species by competitively superior late-successional species (e.g., Glenn-Lewin, Peet, and Veblen 1992; Grime 2001). Besides competition, Connell and Slatyer (1977) mentioned the facilitation of later species by early colonists as an important mechanism of plant succession. Since that time, facilitation has been a vital component of the theory of plant succession (Callaway 1995; Lortie et al. 2004).

Unlike plant community succession, temporal changes in soil microbial communities are not easily detected using visual methods, and more advanced approaches are needed to record the changes (Tscherko et al. 2005). Like different plant species, microbial taxa colonize the soil at different speeds and sequences. As regards succession in mycorrhizal fungal communities, Janos (1980) suggested that mycorrhizal activity may be eliminated from natural ecosystems following disturbance, and then increase during succession. The successional shift from nonmycorrhizal to mycorrhizal plant species may be partly due to the fact that early-successional plant species, which are typically nonmycorrhizal annuals, will be later outcompeted by mycorrhizal perennials. This mechanism was shown to work by Allen and Allen (1984), who demonstrated a reversal in the outcome of competition between early-successional nonmycorrhizal annual *Salsola kali* and two late-successional mycorrhizal perennial grasses, *Agropyron smithii* and *Bouteloua gracilis*. Since late-successional species were only superior competitors in the presence of AM inoculum, the arrival of AMF spores may be crucial for the successional replacement of plant species. Gange, Brown, and Sinclair (1993) showed that the species composition and diversity of early-successional plant communities may differ remarkably between the mycorrhizal and nonmycorrhizal situation.

Early-successional stages in severely disturbed areas are remarkably poor in AMF spores (Allen 1991). The abundance and taxon richness of mycorrhizal fungi in successional ecosystems increases over time due to the stepwise migration of fungal diaspores (Allen et al. 1987; Warner, Allen, and MacMahon 1987; Dodd et al.

2002). A comparison of AMF spore communities along a successional series of old-field ecosystems showed that changes in AMF communities were comparable to vegetation succession—the relative abundance of the spores of eight AM species was clearly different between different successional stages in Cedar Creek, Minnesota. Of the 25 AMF species identified, 7 species were clearly early successional and 5 were late successional (Johnson et al. 1991).

There may be two potential mechanisms through which the facilitation of late-successional species by early-successional ones may be related to the absence or presence of mycorrhizal fungi in the ecosystem. First, the colonization speed and success of plants during succession is dependent not only on the availability of diaspores of plants, but also on the presence of AM fungi (Titus and del Moral 1998; Corkidi and Rincón 1997). Early plant colonists on bare ground or in severely disturbed communities are not very often mycorrhizal-dependent species, since the diaspores of AMF do not possess very efficient dispersal ability; it takes them 25–30 years to recover after disturbance (Boerner, Demars, and Leicht 1996; Hedlund et al. 2004). However, very often the pioneer species are facultatively mycorrhizal, which means that arriving AMF can colonize their roots and, further, make the local biotic and abiotic environment more suitable for the late-successional, more mycotrophic plant species. The early-successional facultative mycotrophic plant species support the qualitative and quantitative accumulation of AMF in the local environment (Corkidi and Rincón 1997), and thereby facilitate the establishment and development of the mycorrhizal-dependent late-successional plant community.

Second, mycorrhizal early colonists may render abiotic conditions more favorable for late-successional plants more efficiently than nonmycorrhizal colonists. Although the nonnutritional effects of mycorrhizae are often overlooked, the extramatrical mycelium (ERM) of AMF is an important binding agent and plays a fundamental role in soil conservation. The ERM of AMF can improve the structure of the soil through the formation of stable soil aggregates by physical entanglement and the production of binding agents that increase its resistance to erosion (Dodd et al. 2002). One of the compounds produced by AM fungi is glomalin, a recalcitrant glycoprotein. In addition to containing substantial carbon, glomalin enhances soil aggregation, thereby protecting carbonaceous material from rapid degradation in soils (Rillig 2004).

5.4.2 THE SPATIAL PATTERN OF AM FUNGAL COMMUNITIES

In principle, differences in mycorrhizal fungal communities over space may result in different species compositions and spatial structures of plant communities, but the potential structuring role of fungal communities depends on how patchy the natural distribution of those fungi is. When the distribution of fungi is more or less uniform, either within one stand or over different stands occurring in similar or different habitat types, they will hardly create any differences within or between plant communities. In contrast, in the case of patchy distribution of mycorrhizal fungi, one may expect a differentiating effect on plant communities as well.

Descriptive studies have shown that natural AM fungal communities vary over space and time to a great extent. First, there may be differences in AMF communities between habitats with different land use. Management practices such as tillage, crop rotation, and fallowing may adversely affect populations of mycorrhizal fungi in the field. Helgason et al. (1998) studied the genetic diversity of root-colonizing AM fungi with the help of molecular markers (SSU rDNA), and compared the fungal communities in herbaceous plant roots in arable fields and in broad-leaved deciduous forest in eastern England. In arable sites, 92% of fungal DNA sequences represented one species, *Glomus mosseae* or closely related taxa, whereas AMF communities from woodlands were much more diverse. Different root-colonizing AMF communities may also characterize different types of natural vegetation. Öpik et al. (2003) recorded different AMF community taxon compositions in boreal grassland and forest. Helgason, Fitter, and Young (1999) found that root-colonizing fungi were different when two forest communities in eastern England—an oak forest and a maple forest—were compared. As concerns the variation of AMF communities within one stand, a considerable small-scale diversity of AMF within a temperate grassland community has been reported (Vandenkoornhuyse et al. 2002).

In a global survey, Öpik et al. (2006) analyzed the results of 26 publications that report on the occurrence of natural root-colonizing AM fungi identified using rDNA region sequences and providing the data that enable a comparison of AM fungal taxon richness and community composition across 36 host plant species and 25 locations. It was found that AM fungal communities exhibit certain similarity in taxon composition within broadly defined habitat types—tropical forests, temperate forests, and habitats under anthropogenic influence. Grassland habitats around the world host heterogeneous AM fungal communities.

These data confirm that AM fungal taxa are patchily distributed in natural and anthropogenic ecosystems. In different ecosystems and even within the same stand, plant individuals may be "exposed" to different symbiotic fungal genotypes. Since the possible indirect and direct effects of AMF taxa on plant–plant interactions vary widely, one may expect a rather complicated pattern resulting from multitrophic multispecies interactions.

5.4.3 INVASION

There are at least three possible ways that mycorrhizal symbiosis can support biological invasions. First, an invasive plant may be positively influenced by the resident AMF assembly, resulting in suppression of indigenous plant species. Second, an invasive AM fungus will be efficiently proliferated by the resident plant community member(s), creating an opportunity for an invasive plant species to match its common symbiont. Third, an arriving alien plant species may carry on its own fungal partners and make the local fungal community more suitable for the next invasions.

There is some evidence supporting the first mechanism. Marler, Zabinski, and Callaway (1999) demonstrated that the native resident microbial community enhanced the negative effect of the invasive weed *Centaurea maculosa* on native *Festuca idahoensis*, and the positive effect was still pronounced when seedlings of the invasive plants grew together with the adults of the native plants. In other experiments, it was found that the competitive dominance of the invasive *Centaurea melitensis* over the native *Nassella pulchra* was sharply reduced when AM fungi in the soil were reduced (Callaway et al. 2001, 2004). The native AMF communities respond to the exotic *C. maculosa* invasion with changes in community composition and abundance: the reduction of diversity and a 25% reduction in the length of the extramatrical mycelium (Mummey and Rillig 2006).

There is not much knowledge about the large-scale transfer of AMF taxa by either natural or anthropogenic vectors (Schwartz et al. 2006). Therefore, the evidence supporting the second or the third mechanism is weak. Since production of commercial mycorrhizal inocula has developed remarkably over recent decades, we may soon hear about invasive AM fungi (Schwartz et al. 2006). The same concerns ectomycorrhizal fungi, which are commercially produced for the inoculation of alien pine species in plantations, while 19 species of *Pinus* have become strongly invasive in the southern hemisphere (Richardson et al. 2000; Schwartz et al. 2006). Though there is not very much empirical evidence, the mutualistic effects of soil biota such as mycorrhizal association are believed to facilitate the invasions (Reinhart and Callaway 2006).

5.4.4 Restoration

It has been recognized for now that the restoration of plant communities also requires an understanding of the functioning of natural mycorrhizal communities (Renker et al. 2004). As described in Section 5.4.1, the presence or absence of mycorrhizal fungi may effect the replacement of plant species and the transformation of soil conditions during primary or secondary succession. Also, mycorrhizal fungi have been shown to possess the ability to accelerate bioremediation (Joner and Leyval 2003).

Plant interactions and mycorrhizal symbiosis are important processes to consider in restoration. Some plant species improve their own environmental conditions by creating "islands of fertility" that facilitate the growth of associated plant species in the vicinity. This phenomenon is common in semiarid communities (e.g., Pugnaire, Haase, and Puigdefabregas 1996; Padilla and Pugnaire 2006). In addition to capturing nutrients and moisture, these islands may trap and proliferate mycorrhizal propagules. It has been documented that endemic *Mimosa* species in a semi-arid community in Mexico (Padilla and Pugnaire 2006) and scrub species in semi-arid areas in southern Spain (Azcon-Aguilar et al. 2003) may serve as mycorrhizal "resource islands" by directly affecting AMF spore dynamics and/or by serving as spore traps.

Since large-scale restoration activities are resource-intensive, one may reduce the cost of restoration by creating resource islands, serving as a source of propagules for plants, mutualistic fungi, and other organisms that may be crucial to an overall development of the ecosystem (Allen, Allen, and Gomez-Pompa 2005). These patches may also be seen as places where mycorrhizal nurse plants create suitable habitats for a range of less-mycorrhizal natural species (Carrillo-Garcia et al. 1999), where transplanted plants create the environment for proliferating mycorrhizal fungi (Rosales et al. 1997; Cuenca, De Andrade, and Escalante 1998), or where pollution-tolerant plant–fungal associations will act and clean the environment by bioremediation (Joner and Leyval 2003) for the less pollution-tolerant successors.

5.5 CONCLUSIONS

Plant interactions and mycorrhizal symbiosis are both important processes structuring natural and anthropogenic plant communities. Mycorrhizae may influence plant interactions either directly by changing plant traits, or indirectly by influencing relationships between plants and organisms of other trophic levels (Figure 5.2).

We can currently conclude that in experimental systems, the presence of AM tends to amplify intraspecific competition and balance interspecific competition. The balancing effect of AM is more evident when plant individuals of different developmental stages compete. However, because the number of experimental studies is still low, the generality of this finding may be challenged due to the biased selection of experimental plant species and AM taxa. This bias might also account for the lack of positive plant interactions (facilitation) in the studies cited.

At the same time, obvious gaps remain in our knowledge. With respect to plant–plant interactions, most experiments so far have compared mycorrhizal and non-mycorrhizal treatments. Such a black-and-white approach might be relevant in the context of the study of ecosystem succession, but its wider significance remains questionable, since most vascular plants in natural communities are mycorrhizal. Another obvious gap is the almost complete lack of knowledge on the effect of AM on plant–plant interactions via multitrophic relationships. We listed some possible hypothetical mechanisms how AM may interfere with plant competition and facilitation via multitrophic relationships, while the next step needs to be experiments to support any of the mechanisms described in this chapter.

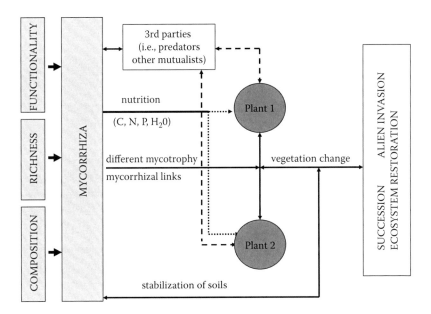

FIGURE 5.2 Simplified scheme showing how mycorrhizal fungi can influence plant individuals and plant–plant interactions and induce the changes in vegetation patterns.

These effects may have a significant outcome on the level of the whole plant community or ecosystem (Figure 5.2) and need further attention.

ACKNOWLEDGMENTS

This study was supported by ETF7366, ETF7371, and SF0180098s08 and by the European Union through the European Regional Development Fund (Center of Excellence FIBIR).

REFERENCES

Allen, E. B., and M. F. Allen. 1984. Competition between plants of different successional stages: Mycorrhizae as regulators. *Can. J. Bot.* 62: 2625–2629.
Allen, E. B., and M. F. Allen. 1990. The mediation of competition by mycorrhizae in successional and patchy environments. In *Perspectives on plant competition*, ed. J. R. Grace, 367–389. New York: Academic Press.
Allen, E. B., J. C. Chambers, K. F. Connor, M. F. Allen, and R. W. Brown. 1987. Natural reestablishment of mycorrhizae in disturbed alpine ecosystems. *Arctic and Alpine Research* 19: 11–20.
Allen, M. F. 1991. *The ecology of mycorrhizae*. Oxford, UK: Cambridge University Press.
Allen, M. F., E. B. Allen, and A. Gomez-Pompa. 2005. Effects of mycorrhizae and nontarget organisms on restoration of a seasonal tropical forest in Quintana Roo, Mexico: Factors limiting tree establishment. *Restor. Ecology* 13: 325–333.
Allsopp, N., and W. D. Stock. 1992. Density dependent interactions between VA mycorrhizal fungi and even-aged seedlings of two perennial Fabaceae species. *Oecologia* 91: 281–287.
Artursson, V., R. D. Finlay, and J. K. Jansson. 2006. Interactions between arbuscular mycorrhizal fungi and bacteria and their potential for stimulating plant growth. *Environ. Microbiol.* 8: 1–10.
Ayres, R. L., A. C. Gange, and D. M. Aplin. 2006. Interactions between arbuscular mycorrhizal fungi and interspecific competition affect size, and size inequality, of *Plantago lanceolata* L. *J. Ecology* 94: 285–294.
Azcon-Aguilar, C., J. Palenzuela, A. Roldan, S. Bautista, R. Vallejo, and J. M. Barea. 2003. Analysis of the mycorrhizal potential in the rhizosphere of representative plant species from desertification-threatened Mediterranean shrublands. *Appl. Soil Ecol.* 22: 29–37.
Barea, J. M., R. Azcon, and C. Azcon-Aguilar. 2002. Mycorhizosphere interactions to improve plant fitness and soil quality. *Antonie Van Leeuwenhoek* 81: 343–351.
Barea, J. M., C. Azcon-Aguilar, and R. Azcon. 1997. Interactions between mycorrhizal fungi and rhizosphere microorganisms within the context of sustainable soil-plant systems. In *Multitrophic interactions in terrestrial systems*, ed. A. C. Gange and V. K. Brown, 65–77. Oxford, UK: Blackwell Science.
Boerner, R. E. J., B. G. Demars, and P. N. Leicht. 1996. Spatial patterns of mycorrhizal infectiveness of soils along a successional chronosequence. *Mycorrhiza* 6: 79–90.
Bonis, A., P. J. Grubb, and D. A. Coomes. 1997. Requirements of gap-demanding species in chalk grassland: Reduction of root competition versus nutrient-enrichment by animals. *J. Ecology* 85: 625–633.
Borowicz, V. A. 2001. Do arbuscular mycorrhizal fungi alter plant-pathogen relations? *Ecology* 82: 3057–3068.

Cahill, J. F., E. Elle, G. R. Smith, and B. H. Shore. 2008. Disruption of a belowground mutualism alters interactions between plants and their floral visitors. *Ecology* 89: 1791–1801.

Callaway, R. M. 1995. Positive interactions among plants. *The Botanical Review* 61: 306–349.

Callaway, R. M., B. E. Mahall, C. Wicks, J. Pankey, and C. Zabinski. 2003. Soil fungi and the effects of an invasive forb on grasses: Neighbor identity matters. *Ecology* 84: 129–135.

Callaway, R. M., B. Newingham, C. A. Zabinski, and B. E. Mahall. 2001. Compensatory growth and competitive ability of an invasive weed are enhanced by soil fungi and native neighbours. *Ecol. Letters* 4: 429–433.

Callaway, R. M., G. C. Thelen, S. Barth, B. W. Ramsey, and J. E. Gannon. 2004. Soil fungi alter interactions between the invader *Centaurea maculosa* and North American natives. *Ecology* 85: 1062–1071.

Carey, E. V., M. J. Marler, and R. M. Callaway. 2004. Mycorrhizae transfer carbon from a native grass to an invasive weed: Evidence from stable isotopes and physiology. *Plant Ecology* 172: 133–141.

Carrillo-Garcia, A., J. L. Leon de la Luz, Y. Bashan, and G. J. Bethlenfalvay. 1999. Nurse plants, mycorrhizae, and plant establishment in a disturbed area of the Sonora Desert. *Restor. Ecology* 7: 321–335.

Colditz, F., H.-P. Braun, C. Jacquet, K. Niehaus, and F. Krajinski. 2005. Proteomic profiling unravels insights into the molecular background underlying increased *Aphanomyces euteichestolerance* of *Medicago truncatula*. *Plant Molecular Biology* 59: 387–406.

Connell, J. H., and R. O. Slatyer. 1977. Mechanisms of succession in natural communities and their role in community stability and organization. *American Naturalist* 111: 1119–1144.

Corkidi, L., and E. Rincón. 1997. *Arbuscular mycorrhizae* in a tropical sand dune ecosystem on the Gulf of Mexico, I: Mycorrhizal status and inoculum potential along a successional gradient. *Mycorrhiza* 7: 9–15.

Corkidi, L., D. Rowland, N. C. Johnson, and E. B. Allen. 2002. Nitrogen fertilization alters the functioning of arbuscular mycorrhizas at two semiarid grasslands. *Plant and Soil* 240: 399–310.

Cuenca, G., Z. De Andrade, and A. E. Escalante. 1998. Arbuscular mycorrhizae in the rehabilitation of fragile degraded tropical lands. *Biological Fertility of Soils* 26: 107–111.

Dickie, I. A., S. A. Schnitzer, P. B., Reich, and S. E. Hobbie. 2005. Spatially disjunct effects of co-occurring competition and facilitation. *Ecol. Letters* 8: 1191–1200.

Dodd, J. C., T. A. Dougall, J. P. Clapp, and P. Jeffries. 2002. The role of arbuscular mycorrhizal fungi in plant community establishment at Samphire Hoe, Kent, UK: The reclamation platform created during the building of the Channel tunnel between France and the UK. *Biodiversity and Conservation* 11: 39–58.

Eissenstat, D. M., and E. I. Newman. 1990. Seedling establishment near large plants: Effects of vesicular-arbuscular mycorrhizas on the intensity of plant competition. *Funct. Ecology* 4: 95–99.

Endlweber, K., and S. Scheu. 2007. Interactions between mycorrhizal fungi and Collembola: Effects on root structure of competing plant species. *Biol. Fert. Soils* 43: 741–749.

Fitter, A. H. 1977. Influence of mycorrhizal infection on competition for phosphorus and potassium by two grasses. *The New Phytologist* 79: 19–125.

Fitter, A. H., and J. Garbaye. 1994. Interactions between mycorrhizal fungi and other soil organisms. *Plant and Soil* 159: 123–132.

Frank, D. A., C. A. Gehring, L. Machut, and M. Phillips. 2003. Soil community composition and the regulation of grazed temperate grassland. *Oecologia* 137, 603–609.

Gange, A. C., V. K. Brown, and G. S. Sinclair. 1993. Vesicular-arbuscular mycorrhizal fungi: A determinant of plant community structure in early succession. *Funct. Ecology* 7: 616–622.

Gange, A. C., and A. K. Smith. 2005. Arbuscular mycorrhizal fungi influence visitation rates of pollinating insects. *Ecol. Entomol.* 30: 600–606.

Gange, A. C., and H. M. West. 1994. Interactions between arbuscular mycorrhizal fungi and foliar-feeding insects in *Plantago lanceolata* L. *New Phytol.* 128: 79–87.

Glenn-Lewin, D. C., R. K. Peet, and T. T. Veblen. 1992. *Plant succession: Theory and prediction.* London: Chapman & Hall.

Goverde, M., M. G. A. van der Heijden, A. Wiemken, I. R., Sanders, and A. Erhardt. 2000. Arbuscular mycorrhizal fungi influence life history traits of a lepidopteran butterfly. *Oecologia* 125: 362–369.

Grime, J. P. 2001. *Plant strategies, vegetation processes, and ecosystem properties.* New York: J. Wiley.

Hart, M. M., R. J. Reader, and J. N. Klironomos. 2003. Plant coexistence mediated by arbuscular mycorrhizal fungi. *Trends in Ecology & Evolution* 18: 418–423.

Hartnett, D. C., B. A. D. Hetrick, G. W. T. Wilson, and D. J. Gibson. 1993. Mycorrhizal influence on intra- and interspecific neighbour interactions among co-occurring prairie grasses. *J. Ecology* 81: 787–795.

Hedlund, K., B. Griffiths, S. Christensen, S. Scheu, H. Setälä, T. Tscharntke, and H. Verhoef. 2004. Trophic interactions in changing landscapes: Responses of soil food webs. *Basic and Applied Ecology* 5: 495–503.

Helgason, T., T. J. Daniell, R. Husband, A. Fitter, and J. P. W. Young. 1998. Ploughing up the wood-wide web? *Nature* 394: 431.

Helgason, T., A. H. Fitter, and J. P. W. Young. 1999. Molecular diversity of arbuscular mycorrhizal fungi colonising *Hyacinthoides non-scripta* (bluebell) in a seminatural woodland. *Mol. Ecol.* 8: 659–666.

Hodge, A., C. D. Campbell, and A. Fitter. 2001. An arbuscular mycorrhizal fungus accelerates decomposition and acquires nitrogen directly from organic material. *Nature* 413: 297–299.

Jaizme-Vega, M., P. Tenoury, J. Pinochet, and M. Jaumot. 1997. Interactions between the root-knot nematode *Meloidogyne incognita* and *Glomus mosseae* in banana. *Plant and Soil* 196: 27–35.

Janos, D. P. 1980. Mycorrhizae influence tropical succession. *Biotropica* 12: 56–64.

Janos, D. P., C. T. Sahley, and L. H. Emmons. 1995. Rodent dispersal of vesicular-arbuscular mycorrhizal fungi in Amazonian Peru. *Ecology* 76: 1852–1858.

Johnson, N. C., D. R. Zak, D. Tilman, and F. L. Pfleger. 1991. Dynamics of vesicular-arbuscular mycorrhizae during old field succession. *Oecologia* 86: 349–358.

Joner, E. J., and C. Leyval. 2003. Phytoremediation of organic pollutants using mycorrhizal plants: A new aspect of rhizosphere interactions. *Agronomie* 23: 495–502.

Kiers, E. T., C. E. Lovelock, E. L. Krueger, and E. A. Herre. 2000. Differential effects of tropical arbuscular mycorrhizal fungal inocula on root colonization and tree seedling growth: Implications for tropical forest diversity. *Ecol. Letters* 3: 106–113.

Kjøller, R., and S. Rosendahl. 1996. The presence of the arbuscular mycorrhizal fungus *Glomus intraradices* influences enzymatic activities of the root pathogen *Aphanomyces euteiches* in pea roots. *Mycorrhiza* 6: 487–491.

Klironomos, J. N. 2003. Variation in plant response to native and exotic arbuscular mycorrhizal fungi. *Ecology* 84: 2292–2301.

Koide, R. T., and I. A. Dickie. 2002. Effects of mycorrhizal fungi on plant populations. *Plant and Soil* 244: 307–317.

Kytöviita, M.-M., M. Vestberg, and J. Tuomi. 2003. A test of mutual aid in common mycorrhizal networks: Established vegetation negates benefit in seedlings. *Ecology* 84: 898–906.

Landis, F. C., A. Gargas, and T. J. Givinish. 2005. The influence of arbuscular mycorrhizae and light on Wisconsin (USA) sand savanna understories, 2: Plant competition. *Mycorrhiza* 15: 555–562.

Legendre, P., and L. Legendre. 1998. *Numerical ecology*. New York: Elsevier.

Lortie, C. J., R. W. Brooker, P. Choler, Z. Kikvidze, R. Michalet, F. I. Pugnaire, and R. Callaway. 2004. Rethinking plant community theory. *Oikos* 107: 433–438.

Maherali, H., and J. N. Klironomos. 2007. Influence of phylogeny on fungal community assembly and ecosystem functioning. *Science* 316: 1746–1748.

Mangan, S. A., and G. H. Adler. 2002. Seasonal dispersal of arbuscular mycorrhizal fungi by spiny rats in a neotropical forest. *Oecologia* 131: 587–591.

Marler, M. J., C. A. Zabinski, and R. M. Callaway. 1999. Mycorrhizae indirectly enhance competitive effects of an invasive forbs on a native bunchgrass. *Ecology* 80: 1180–1186.

Moora, M., M. Öpik, R. Sen, and M. Zobel. 2004. Native arbuscular mycorrhizal fungal communities differentially influence the seedling performance of rare and common *Pulsatilla* species. *Funct. Ecology* 18: 554–562.

Moora, M., M. Öpik, and M. Zobel. 2004. Performance of two *Centaurea* species in response to different root-associated microbial communities and to alterations in nutrient availability. *Annales Botanici Fennici* 41: 263–271.

Moora, M., and M. Zobel. 1996. Effect of arbuscular mycorrhiza on inter- and intraspecific competition of two grassland species. *Oecologia* 108: 79–84.

Moora, M., and M. Zobel. 1998. Can arbuscular mycorrhiza change the effect of root competition between conspecific plants of different ages? *Can. J. Bot.* 76: 613–619.

Mummey, D. L., and M. C. Rillig. 2006. The invasive plant species *Centaurea maculosa* alters arbuscular mycorrhizal fungal communities in the field. *Plant and Soil* 288: 81–90.

Munkvold, L., R. Kjøller, M. Vestberg, S. Rosendahl, and I. Jakobsen. 2004. High functional diversity within species of arbuscular mycorrhizal fungi. *New Phytol.* 164: 357–364.

O'Neill, E. G., R. V. O'Neill, and R. J. Norby. 1991. Hierarchy theory as a guide to mycorrhizal research on large-scale problems. *Environmental Pollution* 73: 271–284.

Öpik, M., M. Moora, J. Liira, U. Kõljalg, M. Zobel, and R. Sen. 2003. Divergent arbuscular mycorrhizal fungal communities colonize roots of *Pulsatilla* spp. in boreal Scots pine forest and grassland soils. *New Phytol.* 160: 581–593.

Öpik, M., M. Moora, J. Liira, and M. Zobel. 2006. Composition of root-colonising arbuscular mycorrhizal fungal communities in different ecosystems around the globe. *J. Ecology* 94: 778–790.

Padilla, F. M., and F. I. Pugnaire. 2006. The role of nurse plants in the restoration of degraded environments. *Frontiers in Ecology and the Environment* 4: 196–202.

Pietikainen, A., and M. M. Kytöviita. 2007. Defoliation changes mycorrhizal benefit and competitive interactions between seedlings and adult plants. *J. Ecology* 95: 639–647.

Pugnaire, F. I., P. Haase, and J. Puigdefabregas. 1996. Facilitation between higher plant species in a semiarid environment. *Ecology* 77: 1420–1427.

Read, D. 1997. The ties that bind. *Nature* 388: 517–518.

Redecker, D., J. B. Morton, and T. D. Bruns. 2000. Ancestral lineages of arbuscular mycorrhizal fungi (Glomales). *Molecular Phylogenetics and Evolution* 14: 276–284.

Reinhart, K. O., and R. M. Callaway. 2006. Soil biota and invasive plants. *New Phytol.* 170: 445–457.

Renker, C., M. Zobel, M. Öpik, M. F. Allen, E. B. Allen, M. Vosatka, J. Rydlova, and F. Buscot. 2004. Structure, dynamics and restoration of plant communities: Do arbuscular mycorrhizae matter? In *Assembly rules and restoration ecology: Bridging the gap between theory and practice*, ed. V. Temperton, R. Hobbs, T. Nuttle, and S. Halle, 189–229. Washington, DC: Island Press.

Richardson, D. M., N. Allsopp, C. M. D'Antonio, S. J. Milton, and M. Rejmanek. 2000. Plant invasions: The role of mutualism. *Biological Review* 75: 65–93.

Rillig, M. C. 2004. Arbuscular mycorrhizae and terrestrial ecosystem processes. *Ecol. Letters* 7: 740–754.

Robinson, D., and A. Fitter. 1999. The magnitude and control of carbon transfer between plants linked by a common mycorrhizal network. *J. Exp. Bot.* 50: 9–13.

Ronsheim, M. L., and S. E. Anderson. 2001. Population-level specificity in the plant-mycorrhizae association alters intraspecific interactions among neighboring plants. *Oecologia* 128: 77–84.

Rosales, J., G. Cuenca, N. Ramirez, and Z. De Andrade. 1997. Native colonizing species and degraded land restoration in La Grand Sabana, Venezuela. *Restor. Ecology* 5: 147–155.

Rosendahl, S. 2008. Communities, populations and individuals of arbuscular mycorrhizal fungi. *New Phytol.* 178: 253–266.

Scheublin, T. R., R. S. P. van Logtestijn, and M. G. A. van der Heijden. 2007. Presence and identity of arbuscular mycorrhizal fungi influence competitive interactions between plant species. *J. Ecology* 95: 631–638.

Schussler, A., D. Schwarzott, and C. Walker. 2001. A new fungal phylum, the Glomeromycota: Phylogeny and evolution. *Mycological Research* 105: 1413–1421.

Schwartz, M. W., J. D. Hoeksema, C. A. Gehring, C. N. Johnson, J. N. Klironomos, L. K. Abbott, and A. Pringle. 2006. The promise and the potential consequences of the global transport of mycorrhizal fungal inoculum. *Ecol. Letters* 9: 501–515.

Simard, S. W., and D. M. Durall. 2004. Mycorrhizal networks: A review of their extent, function, and importance. *Can. J. Bot.* 82: 1140–1165.

Smith, S. E., and D. J. Read. 1997. *Mycorrhizal symbiosis*. New York: Academic Press.

StatSoft. 2001. STATISTICA for Windows. Tulsa, StatSoft Inc.

Titus, J. H., and R. del Moral. 1998. The role of mycorrhizal fungi and microsites in primary succession on Mount St. Helens. *Am. J. Bot.* 85: 370–375.

Tscherko, D., U. Hammesfahr, G. Zeltner, E. Kandeler, and R. Böcker. 2005. Plant succession and rhizosphere microbial communities in a recently deglaciated alpine terrain. *Basic and Applied Ecology* 6: 367–383.

Vandenkoornhuyse, P., R. Husband, T. J. Daniell, I. J. Watson, J. M. Duck, A. H. Fitter, and J. P. W. Young. 2002. Arbuscular mycorrhizal community composition associated with two plant species in a grassland ecosystem. *Mol. Ecol.* 11: 1555–1564.

van der Heijden, M. G. A. 2004. Arbuscular mycorrhizal fungi as support systems for seedling establishment in grassland. *Ecol. Letters* 7: 293–303.

van der Heijden, M. G. A., R. D. Bardgett, and N. M. van Straalen. 2008. The unseen majority: Soil microbes as drivers of plant diversity and productivity in terrestrial ecosystems. *Ecol. Letters* 11: 296–310.

van der Heijden, M. G. A., T. Boller, A. Wiemken, and I. R. Sanders. 1998. Different arbuscular mycorrhizal fungal species are potential determinants of plant community structure. *Ecology* 79: 2082–2091.

van der Heijden, M. G. A., R. Streitwolf-Engel, R. Riedl, S. Siegrist, A. Neudecker, K. Ineichen, T. Boller, A. Wiemken, and I. R. Sanders. 2006. The mycorrhizal contribution to plant productivity, plant nutrition and soil structure in experimental grassland. *New Phytol.* 172: 739–752.

van der Heijden, M. G. A., A. Wiemken, and I. R. Sanders. 2003. Different arbuscular mycorrhizal fungi alter coexistence and resource distribution between co-occurring plant. *New Phytol.* 157: 569–578.

Warner, N. J., M. F. Allen, and J. A. MacMahon. 1987. Dispersal agents of vesicular-arbuscular mycorrhizal fungi in a disturbed arid ecosystem. *Mycologia* 79: 721–730.

West, H. M. 1996. Influence of arbuscular mycorrhizal infection on competition between *Holcus lanatus* and *Dactylis glomerata*. *J. Ecology* 84: 429–438.

Wilson, G. W. T., D. C. Hartnett, and C. W. Rice. 2006. Mycorrhizal-mediated phosphorus transfer between tallgrass prairie plants *Sorghastrum nutans* and *Artemisia ludoviciana*. *Funct. Ecology* 20: 427–435.

Wolfe, B. E., B. C. Husband, and J. N. Klironomos. 2005. Effects of a belowground mutualism on an aboveground mutualism. *Ecol. Letters* 8: 218–223.

Zobel, M., and M. Moora. 1995. Interspecific competition and arbuscular mycorrhiza: Importance for the coexistence of two calcareous grassland species. *Folia Geobotanica et Phytotaxonomica* 30: 223–230.

6 Plant Communities, Plant–Plant Interactions, and Climate Change

Rob W. Brooker

CONTENTS

6.1   INTRODUCTION

This chapter aims to demonstrate how consideration of the role of plant–plant interactions has direct relevance to understanding and coping with the impacts of one of the most serious threats to biodiversity that has resulted from the activities of mankind—climate change.

I will first provide a little background information on climate change and our current understanding of likely future changes in climate. Second, I will combine information on the role of plant–plant interactions provided by long-term monitoring studies, experimental manipulations of plant communities, and computer modeling to explore how plant–plant interactions might mediate the impacts of climate change on plant communities. Third, I will discuss the need to predict the future impacts of climate change on plant communities, how inclusion of the role of plant–plant interactions in developing these predictions is essential, and how current developments in plant–plant interaction theory might have direct relevance to providing and improving such predictions. Finally, I will discuss some potentially fruitful areas for future research that might improve both our understanding of the role of plant–plant interactions in plant communities and our ability to predict and manage the impacts of climate change on biodiversity.

I do not intend to provide a comprehensive review of the impacts of climate change on plant–plant interactions, or of the links between plant interactions and climate (discussed in greater detail elsewhere in this book). Instead, my primary aim is to use a few selected studies to illustrate underlying fundamental points that are relevant to all plant communities, irrespective of their location. Furthermore, the examples that I will use have an admittedly European bias. However, evidence for the patterns and processes that I discuss can be found from ecosystems across the globe, and I hope that the concepts discussed here are clear enough to be readily extrapolated beyond the confines of temperate/northern European plant communities.

6.2 CLIMATE CHANGE: A BRIEF OVERVIEW

It is now impossible to ignore the phenomenon of anthropogenic climate change. Although the Earth's climate has always oscillated between warm and cold periods, there is "very high confidence" that human activity is influencing global climatic conditions through anthropogenic climate forcing resulting from release of greenhouse gases into the atmosphere (IPCC 2007). Eleven of the last twelve years are among the twelve warmest years in the instrumental record of global surface temperature, and the linear warming trend for 1956 to 2005 is nearly double that for the 100 years from 1906 to 2005 (IPCC 2007). Fossil fuel combustion and cement production have contributed significantly to increases in the concentration of the main greenhouse gas, CO_2 (carbon dioxide), from a preindustrial level of approximately 270–280 ppm (parts per million) to the current (2008) tropospheric concentration of over 384.8 ppm (Tans 2009). Other important anthropogenic greenhouse gases include CH_4 (methane) from agriculture, N_2O (nitrous oxide) from agriculture and industry, and halogenated gases and ozone from industrial and domestic sources (EEA 2004). Temperature change is not the only component of anthropogenic climate change. Other likely impacts include changes in precipitation and wind patterns and the frequency and some aspects of extreme climatic events (IPCC 2007). The impacts of climate change are not spatially uniform. For example, Europe has warmed more rapidly than the global average temperature, with a 0.95°C increase since 1900. During the same period, northern Europe has experienced a 10%–40% increase in precipitation, while southern Europe has experienced a decrease of up to 20% (EEA 2004). Hereinafter I will use the phrase "climate change" to refer to anthropogenic climate change (as opposed to natural climatic fluctuations).

To develop policies that enable society to adapt, predictions of the future impacts of climate change are essential. However, the development of societies and their associated greenhouse gas emissions are extremely difficult to predict. In addition, we currently have an incomplete understanding of climate processes. Therefore, it is not possible to produce "predictions" of the extent of future climate change (EEA 2004). Instead, GCMs (general circulation models), parameterized with emission scenarios that include the potential responses of society to the threat of climate change (e.g., a significant change in habits compared to "business as usual"), are used to provide future climate "scenarios." Because of variation in the emission scenarios and GCMs, there is considerable variation in the details of the future climate scenarios produced, and averages from multiple runs of a range of different GCMs are used to provide an indication of the likely future climate and the uncertainty associated with these predictions. At a very broad scale, they indicate a continued warming of the global average surface temperature of between 1.8°C and 4.0°C above 1980–1999 levels by 2090–2099 (IPCC 2007).

6.3 CLIMATE CHANGE AND PLANT COMMUNITIES

6.3.1 SOURCES OF INFORMATION

Evidence for the impact of climate change on natural systems comes from three main sources:

1. Monitoring studies, which relate observed changes in the composition of communities or the distribution of species to long-term trends in climatic conditions
2. Experimental manipulations, which alter some component of climate, for example temperature or precipitation, in a (more or less) controlled fashion and then monitor changes in the plant community relative to unmanipulated control plots
3. Modeling studies

All three of these approaches have both positive and negative aspects.

Monitoring studies are of necessity purely correlative, but at the same time, because of their longevity and common focus on attractive flora and fauna, they can provide some of the most persuasive indications of the impacts of climate change, such as upward altitudinal shifts in European alpine plant species and northerly shifts in European bird species (Grabherr, Gottfried, and Pauli. 1994; Thomas and Lennon 1999).

Unlike monitoring, manipulative experimental studies can demonstrate that the response of a particular community is the result of changes in particular climate variables (rather than a possible coincidental concomitant change in community composition and climate). Small-scale plot experiments (Figure 6.1), using a range of experimental approaches, for example using open-topped chambers (OTCs), soil-warming cables, or nighttime covers to enhance temperatures (with greater or lesser realism in terms of creating modeled future climate), have been conducted in a wide range of environments, and have given valuable insights into the possible responses of species, communities, and ecosystem function to climate change (e.g., Arft et al. 1999; Emmett et al. 2004; Klein, Harte, and Zhao. 2004; Morecroft et al. 2004; Walker et al. 2006). However, climate manipulation experiments are commonly undertaken on small plots, making it difficult to extrapolate from the plot to larger

(a)

(b)

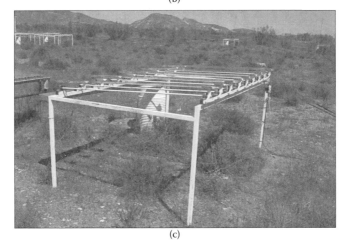

(c)

FIGURE 6.1

scales, creating potential artifacts. For example, community invasion by species from lower latitudes may be one of the most serious threats to arctic plant communities, but this is a process that is not tackled by manipulating temperatures in small plots that do not include potential invaders.

Finally, modeling studies seek to determine the responses of individual species, communities, or ecosystem services (often of key interest to policy makers; Watson 2005) to future changes in climate. A wide range of modeling approaches are applied, including individual-based, mechanistic computer models and the commonly used climate envelope modeling approach, and their output can be more or less abstract. For example, Travis (2002) used individual-based cellular automata modeling to examine the impact of habitat fragmentation on the range-shifting capacity of a hypothetical species within a two-dimensional grid model space (of direct relevance to the ability of species to track climate change). At the other end of the spectrum, Berry et al. (2002) used a correlative bioclimatic envelope modeling approach to assess future changes in the potential range of particular plant species within the United Kingdom.

6.3.2 CLIMATE CHANGE AND COMMUNITY COMPOSITION

Long-term monitoring studies have indicated changes in the composition of plant communities that might be attributable to climate change. For example, in northern temperate European ecosystems, these changes commonly involve a decline in species with a generally northern distribution and an increase in species with a generally southern distribution (Preston et al. 2002; EEA 2004; Parmesan 2006).* These changes in community composition might result from either changes within the extant plant community or the influx of new species. As mentioned, in alpine environments there is well-documented evidence of an upward shift in species

FIGURE 6.1   (Facing Page) (*A color version of this figure follows page 110.*) Typical climate manipulation experiments. At Abisko Research Station in northern Sweden, the experiments use Plexiglas open-topped chambers (a), based on the widespread ITEX design (e.g., Molau and Alatalo 1998; Arft et al. 1999), to enhance air (and soil) temperatures in small patches of low-lying vegetation. The response of the vegetation is assessed by comparing plots within the chambers to untreated control plots (b). The duckboards shown protect the delicate tundra vegetation from trampling. (Photos kindly provided by Hans Cornelissen, Vrije Universiteit, Amsterdam.) In contrast, in the Almeria region of southern Spain, experimental drought treatments are applied by using frames that collect rainwater and divert it away from the plot (c).

---

* Throughout this chapter, I discuss mainly the response of northern temperate, alpine, subarctic, or arctic plant communities, wherein enhanced temperatures will likely lead to enhanced productivity and reduce the degree of direct limitation on growth resulting from the abiotic environment. However, it should be recognized that the opposite pattern of responses may occur in drier, more southerly (e.g., Mediterranean and desert) systems, such that enhanced temperatures increase the direct negative impact of abiotic environmental conditions, for example either directly through temperature stress or indirectly through water limitation. The examples provided here are therefore only illustrative of underlying processes, rather than indicating a lack of response in warmer, lower latitude systems.

distributions that correlates closely with recent temperature increases (Grabherr, Gottfried, and Pauli 1994). Range shifts have also been detected in alpine tree lines (Kullman 2001), in individual European plant species (e.g., holly *Ilex aquifolium*; Walther, Berger, and Sykes 2005), and are clearly demonstrated by the expansion of shrub species into tundra regions (Sturm, Racine, and Tape 2001; Tape, Sturm, and Racine 2006).

Lortie et al. (2004) presented a diagram (similar to those proposed by Grime [1998] and Laakso, Kaitala, and Ranta [2001]) of the range of processes that are involved in regulating the composition of plant communities, and translating the potential global species pool into the realized community at a given point in time and space (Figure 6.2). All of the processes presented in this diagram might be influenced in some way by climate, and thus play a role in determining the response of plant communities to climate change (Table 6.1). Certainly observed responses of plant communities and species would indicate (based on our simple schema) that either the capacity of species to reach more northerly or higher altitude sites has increased, or that they now have an increased chance of survival having reached those sites. I will now discuss in more detail how climate change might be influencing some of these processes, in particular seed production and dispersal, and plant–plant interactions in communities.

6.3.2.1 Seed Production and Dispersal

The dispersal of many species is likely to be influenced by climate, operating through a number of mechanisms. First, propagule availability may increase in systems where it was previously temperature-limited. A classic example of this is the regulatory impact of low temperatures on the production of viable seed by *Tilia cordata* at the northern limits of its UK range; significant numbers of viable seed were only produced during (what were then considered to be) extremely warm summers (Pigott and Huntley 1981). Enhanced temperatures might increase insect activity, seed set following pollination, or the ripening of seed following seed set. The impact of low temperatures on the development of viable seed is extremely important in more northerly latitudes or in montane or alpine ecosystems, where the dominance of clonal reproduction is thought to demonstrate the difficulties of developing ripe seed within the short, cool growing season (Callaghan, Jonasson, and Brooker 1997). Climate manipulation experiments in high arctic environments have shown a strong positive response of seed production to enhanced temperatures (Arft et al. 1999).

Second, the movement of mature propagules might also be regulated by climate, particularly in those species that utilize animal vectors for dispersal; changes in the migratory habits of birds and animals, as well as the patterns of human activity, will impact on the distances and directions over which propagules are dispersed. However, although propagule production and dispersal may increase, the spread of the species is then dependent on the availability of suitable habitat and germination and establishment conditions. Land-use change is leading to significant fragmentation of habitats, may greatly restrict the capacity of species to move through the landscape in response to climate change (Travis 2002), and in some cases may even lead to local extinction due to contracting southern range margins and restricted

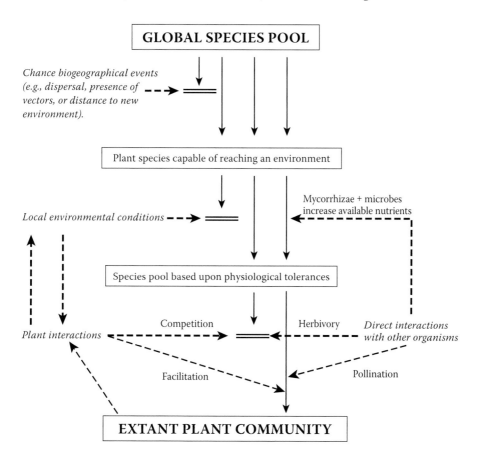

FIGURE 6.2   Simple flow diagram illustrating the main processes (or filters) that structure a plant community, and how they operate to convert the global species pool into the realized extant plant community at a given site. These filtering processes are represented by a pair of horizontal lines, with a corresponding description in bold italics. Solid arrows depict the movement of species through the filters, and hatched lines illustrate where each process might influence the plant community. Climate change may act to influence the impact of all of these filters (see text and Table 6.1), and the role of plant–plant interactions in mediating the impacts of climate change must be viewed within the context of the suite of processes that regulate plant community composition. (Reproduced with permission from Lortie et al. 2004. *Oikos* 107: 433–438.)

expansion of northern range margins (in northern hemisphere species; Honnay et al. 2002).

6.3.2.2   Interactions in Communities

If a species is able to reach suitable habitat (despite the problems of dispersal limitation and habitat fragmentation), the probability of a species establishing is governed by its capacity to tolerate local environmental conditions, which include not only

TABLE 6.1

The Possible Impact of Climate Change on the Filters that Regulate the Composition of Plant Communities (see Figure 6.2)

Filter	Impact of Climate Change	Ecological Response
Biogeographical events: dispersal, presence of vector, distance to new environment	Increase propagule production through reduced temperature limitation, increased pollinator activity, and faster rate of seed maturation Changes in migratory habits of animal vectors (expansion of high latitude/altitude range limits)	Larger number of propagules reaching the region of suitable climatic conditions
Local environmental conditions (abiotic)	Reduction in the severity of abiotic environmental conditions (e.g., enhanced temperatures, increased soil nutrient availability) Changes in phenology (earlier growing season, e.g., through earlier spring warming, earlier snow melt)	Release of constraint on growth for species in situ previously at extreme northern limits of range Influx of new species previously excluded by physiological intolerance of environment Changes in relative competitive dominance within community
Direct interactions with other organisms (herbivores or pollinators)	Changes in level of grazing (possibly increasing with increased productivity) Increased pollination through enhanced insect activity	Promotion of grazing tolerant species, changes in competitive dominance within community Enhanced propagule production and fecundity of sexually reproducing species Change in reproductive balance between sexually and clonally reproducing species
Plant–plant interactions	Changes in relative interaction balance of species within community, e.g., promotion of more competitive species and loss of stress-tolerant facilitators	Competitive exclusion of slower-growing stress-tolerant species

Note: These impacts are discussed within the context of northern temperate, arctic, or alpine plant communities, but it is straightforward to extrapolate similar impacts for warmer, low-latitude systems.

the abiotic environment, but also the biotic environment generated by those species already in situ (Figure 6.2). As discussed in the previous subsection, climatic changes may drive changes in the distribution of herbivorous or pollinator species that will influence the success of plants at a given site. Differences in the ability of species to track climatic conditions are already leading to the development of novel species combinations with unexpected consequences for community composition. For example, Hódar and Zamora (2004) found that increased winter temperatures led to an upward altitudinal shift in the range of the pine processionary moth

Thaumetopoea pityocampa in the Spanish Sierra Nevada, such that its range now overlaps with relic populations of *Pinus sylvestris* ssp. *nevadensis*, which has subsequently suffered from severe defoliation.

However, of particular interest here is the impact of climate change on plant–plant interactions and their regulation of community composition, although the previous examples are important in that they illustrate how interactions are only one of a number of processes that mediate climate-change impacts on communities. For simplicity I will consider two broad categories of plant interactions: competition and facilitation.

6.3.2.3   Competitive Plant–Plant Interactions

A number of processes have been proposed as important in regulating community composition and species diversity (as illustrated in Figure 6.2), including evolutionary history and the available species pool (Zobel 1997; Foster 2001; Foster et al. 2004) and random fluctuations in species abundance (Hubbell 2001; McGill 2003; Nee and Stone 2003). However, it is widely accepted that competitive interactions—including competition for germination and establishment space as well as competitive impacts on growth and reproduction—and the capacity of the environment to limit the impacts of competition (e.g., through disturbance) are critical processes in determining the diversity of species within plant communities (Grime 1973).

Changes in climate can strongly influence the competitive impact of species by imposing or removing physiological limitations on establishment and growth, or by altering factors such as nutrient availability or phenology. Manipulation experiments in arctic and alpine environments have shown that shrubs and graminoids are particularly responsive to enhanced temperatures or nutrient additions (e.g., Press, Callaghan, and Lee 1998; Dormann and Woodin 2002; Brooker and Van der Wal 2003), and enhanced growth of such species and associated increases in the level of competition have been proposed as the cause of changes in community composition (e.g., Harte and Shawe 1995; Dormann, Van der Wal, and Woodin 2004; Heegaard and Vandvik 2004). For example, enhanced competition for light, due to increased vascular plant growth, is thought to be the cause of the general decline in cryptogam species in tundra-based warming experiments (Molau and Alatalo 1998; Cornelisson et al. 2001). In the mid to polar latitudes there is already evidence of a general increase in productivity associated with climate change (Zhou et al. 2001), and the expansion of more "southerly" species—no longer constrained by abiotic environmental conditions, for example shrubs in tundra environments (Sturm, Racine, and Tape 2001, Tape, Sturm, and Racine 2006)—may alter the "competitive climate" within communities. However, although these changes may occur in the extant plant community through shifts in the relative dominance of species, competition for space may limit further changes in community composition by preventing the establishment of more southerly species not already present within a community (Takenaka 2005), thereby creating a "tensioned" landscape (Brooker et al. 2007).

Changes in phenology have been highlighted as one of the most consistent and compelling "fingerprints" of climate change, and have been detected in many groups of organisms, including butterflies, birds, amphibians, and plants (Parmesan and Yohe 2003; Root et al. 2003; Parmesan 2006). A general trend toward earlier bud burst and flowering times has been detected in the European flora (Sparks and Menzel 2002; Walther et al. 2002), and may result from an earlier onset of the spring warming period (Schwartz, Ahas, and Aasa 2006). Avoidance of competition is considered to be an important driving force in the evolution of temporal niches and the patterns of phenology that occur during the growing season; a classic example is the early flowering of woodland understory species, such as the bluebell *Hyacinthoides non-scripta*, in order to avoid light limitation from the tree canopy. Changes in phenological processes can therefore influence the relative competitive balance of species within communities, especially as there appears to be species-specific variation in phenological responses to climate change (Peñuelas and Filella 2001; Walther 2003). Early flowering plants have been found to be particularly responsive to a shift toward an earlier onset of the growing season, and the level of response is also dependent upon the position of individuals within the overall range of the species (Fitter and Fitter 2002). Earlier growth might enable some species to monopolize more resources, thus further increasing their positive response to climate change. Dunnett and Grime (1999) proposed that "climatic impacts will be most strongly amplified by interspecific competition" in more productive environments as competition plays a greater role in determining community composition (Grime 1979). Dunnett and Grime demonstrated that five common UK plants, when grown in monocultures, all responded to enhanced spring temperatures with earlier growth. However, in mixture only a subset of species showed earlier phenology, indicating that competition can amplify the direct impacts of climate on community composition.

6.3.2.4 Facilitative Plant–Plant Interactions

As discussed elsewhere in this book, although the role of competition in regulating the composition of plant communities has been a widely accepted component of plant community theory for many years, the role of facilitation has only recently received particular attention (see also Bertness and Callaway 1994; Callaway 1995; Brooker and Callaghan 1998; Callaway and Pugnaire 1999; Bruno, Stachowicz, and Bertness 2003; Brooker et al. 2008). With respect to climate change, few studies have explicitly considered the role of facilitation in regulating the response of community composition, perhaps because of the dominance of competition in plant community ecology. Despite this, it is possible to infer the role that facilitation might play during climate change.

In brief, facilitation occurs when the presence of one plant is able to ameliorate the environmental conditions that are limiting the growth of a neighboring plant. In some cases these limiting factors might be biotic disturbance processes, for example grazing (e.g., Rousset and Lepart 1999, 2000; Bellingham and Coomes 2003; Rao et al. 2003; Brooker et al. 2006). In many cases, however, the limiting factors are the abiotic components of the plants' environment. For example, in arctic and alpine ecosystems, neighboring plants can provide protection from ice crystal abrasion and water loss, limit frost heave damage to root systems (by insulating the soil), accumulate leaf litter leading to enhanced soil nutrient availabilities, and raise leaf tissue temperatures through the provision of a favorable microclimate (Brooker et al. 2008). In salt marsh systems, those species with a higher salt tolerance can promote the growth of less salt-tolerant species by limiting the evaporative loss of water from the soil and thus the accumulation of salt (Bertness and Callaway 1994). In desert

systems, nurse plants accumulate organic matter, enhancing soil moisture and nutrient availability, as well as providing areas of shade (and reduced temperatures) that
are commonly utilized as microsites for the germination and growth of desert annuals (Callaway and Pugnaire 1999). It is these processes, related to amelioration of
abiotic environmental conditions (and thus the prevailing climate at the site), that are
perhaps of the greatest interest when considering the response of plant communities
to climate change.

There is clear evidence that the impact of facilitation, as with competition, is
regulated by variation in climatic conditions, and this demonstrates that climate
change is likely to influence the role of facilitation within plant communities. Recent
studies in alpine ecosystems have shown that the dominant type of interaction within
plant communities (e.g., competition or facilitation) is related to the prevailing climatic conditions or the position of the target species within its own range (effectively
its position along a climatic gradient relative to some nominal species optimum).
Callaway et al. (2002) demonstrated a large-scale pattern in the dominant type of
interaction within alpine plant communities: An increase in the productivity of the
environment (as indicated by the average June temperature) was associated with
a general switch from facilitation to competition. Choler, Michalet, and Callaway
(2001) found a similar switch from competition to facilitation with increasing environmental severity (altitude) in French alpine systems, and also that the net type
of interaction experienced by the target plant is dependent on the position of the
individual within the overall species' range. They state, "When neighbours were
removed from around target species at experimental sites that were lower in elevation than the distributional mean of the target species, biomass generally increased
[indicating competition]. When neighbours were removed from around target species at experimental sites that were higher in elevation than the distributional mean
of the target species, biomass generally decreased [indicating facilitation]." This
demonstrates that the impact of interactions might not only be growth-form specific
(as discussed in Section 6.3.2.3 with respect to the significant positive response of
arctic/alpine graminoids and shrubs to experimental warming), but might also be
location specific.

Similar patterns of the relationship between facilitative interactions and climate
have been found for salt-marsh ecosystems. Bertness and Ewanchuk (2002) examined facilitative plant–plant interactions in salt-marsh communities in Rhode Island
and Maine. They found a greater impact of facilitation at the warmer of the two sites
(Rhode Island) and during warmer years. In these ecosystems, high temperatures
equate with water limitation and increased soil salinity, and the matrix species (the
common facilitators) shade and cool the soil. Again the interaction experienced was
growth-form specific: Salt-intolerant species tended to benefit most from the presence of neighbors, whereas salt-tolerant species (which would not reap the benefits of
facilitation) tended to suffer from competition with the matrix species.

Although very few climate change manipulation experiments have explicitly
considered the role of facilitation, that conducted by Klanderud (2005) was specifically aimed at considering the possible switch from facilitation to competition that
might occur for some species in alpine or arctic ecosystems during climate change.
The study found evidence of increased competition following warming or nutrient

addition, but was less conclusive about the role and change in the level of facilitation that occurred following the experimental manipulations. For example, removal of the *Dryas octopetala* vegetation matrix led to increased *Thalictrum alpinum* and *Carex vaginata* leaf production (indicating competition from the *Dryas*), but also to reduced length of flowering stalk and leaves (which may indicate either a loss of facilitative effects or a loss of competition-induced etiolation). However, the study is right to conclude that interactions play an important role in regulating the response of the plant communities to experimental manipulations, and should be considered in more detail by future community manipulation experiments. Such experiments should also consider in more detail the role of wind speed. The type of OTC used by Klanderud and many other studies of the impacts of temperature on arctic and alpine plant communities (e.g., Arft et al. 1999; Walker et al. 2006) manipulates temperature by both trapping infrared radiation (in the manner of a greenhouse) and by blocking wind and hence limiting forced convective heat loss from the vegetation as well as mechanical damage. Future predicted changes in temperature, with enhanced temperatures in alpine environments improving growing conditions for plants, are not necessarily matched by changes in wind speed. It is therefore important for us to understand the relative balance of the direct impact of temperature and the effects of wind through changes in temperature and mechanical damage on the potential upward or poleward movement of species limits.

However, based on the available evidence from existing climate-change studies, we might tentatively make some predictions as to the role of facilitative interactions in mediating the impact of climate change on plant communities. In arctic and alpine systems, the proportional role of facilitative interactions will decrease as the productivity of the environment increases. In terms of an impact on plant community composition, the effect may be negligible; in these systems it is likely that those species benefiting most from facilitation are at the severe end of their climatic range (Michalet et al. 2006). Increased productivity in these systems will not lead to a loss of these species, as their range optimum lies within warmer conditions anyway. Mediterranean systems, however, will become increasingly severe, being both hotter and drier (IPCC 2007), and facilitation may play an increasing role in maintaining species that would otherwise be intolerant of the prevailing abiotic environmental conditions. So we see that (as with competition) changes in the role of facilitation under climate change are likely to be context-specific. However, in terms of predicting such changes, much depends upon the shape and generality of the links between climate and the role of plant–plant interactions, an issue that has been the subject of considerable debate and that I will return to later in this chapter.

To briefly summarize the topics covered so far, we have seen how there are many routes by which climate change might impact the composition of plant communities, including the provision and transport of propagules, and the suitability of the environment (both biotic and abiotic) for plant establishment and growth. More specifically we have seen that plant interactions (both facilitation and competition) are capable of playing an important role in determining the response of plant communities to climate change, but that the exact nature of the role (and how it might change in relation to climate) depends upon the species and environment concerned. As I will discuss below, attempting to impose some order on these at-first apparently

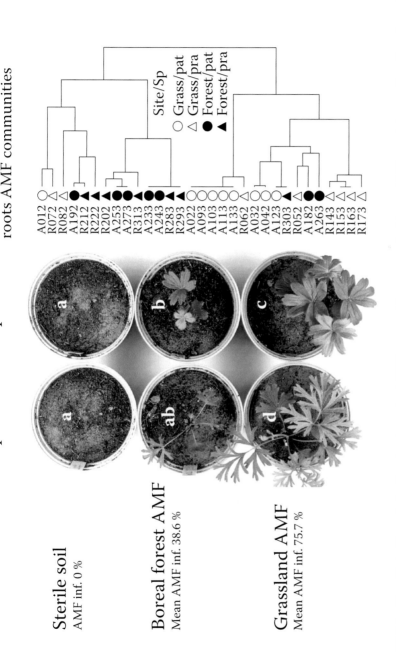

COLOR FIGURE 5.1 The effect of boreal pine forest and grassland AMF communities compared with non-AMF containing sterile soil on the performance of *Pulsatilla pratensis* and *P. patens* after 14 weeks growth in a greenhouse. Different lowercase letters above pots indicate a statistically significant difference in total dry mass. Mean AMF infection is the percent of the colonized root length. The results of hierarchical clustering analysis of AM fungal communities in *Pulsatilla* seedling roots are also presented. AM fungi were identified as SSU rDNA sequence groups by Öpik et al. (2003), and each sample (row in the picture) represents the root system of one seedling where the fungal community is described by fungal sequence types present/absent. *Pulsatilla pratensis* (triangles) and *P. patens* (circles) were grown in the presence of forest (filled symbols) and grassland (open symbols) soil inoculum. Figure is modified by Moora et al. (2004). (Photograph by R. Sen.)

(a)

COLOR FIGURE 6.1 Typical climate manipulation experiments. At Abisko Research Station in northern Sweden, the experiments use Plexiglas open-topped chambers (a), based on the widespread ITEX design (e.g., Molau and Alatalo 1998; Arft et al. 1999), to enhance air (and soil) temperatures in small patches of low-lying vegetation. The response of the vegetation is assessed by comparing plots within the chambers to untreated control plots (b). The duckboards shown protect the delicate tundra vegetation from trampling. (Photos kindly provided by Hans Cornelissen, Vrije Universiteit, Amsterdam.) In contrast, in the Almeria region of southern Spain, experimental drought treatments are applied by using frames that collect rainwater and divert it away from the plot (c).

COLOR FIGURE 6.1 (Continued).

COLOR FIGURE 6.1 (Continued).

random responses may be critical in developing approaches for predicting the future impacts of climate change on the composition of plant communities.

6.4 PREDICTING PLANT COMMUNITY RESPONSES: CONSIDERING INTERACTIONS

Above I have mainly discussed information from monitoring and experimental manipulations. I have not yet discussed the third main source of information—modeling—as it is important to examine in greater detail current debate surrounding the integration of plant–plant interactions (and indeed biotic interactions in general) into the modeling process.

Climate change has been acknowledged as one of the major current threats to global biodiversity (CBD 2003; Millennium Ecosystem Assessment 2005). Some researchers have even suggested that its potential impact may exceed other major anthropogenic drivers of global environmental change (Thomas et al. 2004), although others disagree (Buckley and Roughgarden 2004; Harte et al. 2004; Thuiller et al. 2004). Irrespective, it is now generally acknowledged that policies must be developed in order to maintain and conserve biodiversity during future climate change (CBD 2003; Brooker and Young 2006). Two types of policy are often discussed with respect to managing the impacts of climate change: mitigation policies (limiting the level of climate change, for example through controlling greenhouse gas emissions) and adaptation policies (dealing with the impacts of the changes in global climate that will occur irrespective of the mitigation measures taken). However, the development of such policy commonly necessitates some understanding of the likely impact of climate change; in the case of biodiversity conservation, this means understanding the future impacts of climate change on biodiversity.

Mathematical, statistical, and computer simulation modeling is one of the major tools available for providing this information. Modeling has advantages over field-based manipulations in that models can deal with much larger areas than field manipulations, and so provide information at scales directly relevant to policy development (e.g., regional, national, and international).

One major modeling approach—climate envelope modeling—deals with the impact of climate change on the distribution and abundance of organisms (which ultimately determines the composition of communities). Put simply, this approach statistically determines the relationship between a target species' spatial distribution pattern and a range of explanatory climatic parameters. Based on the model fit, and using future climate scenarios provided by GCMs (general circulation models), the future spatial distribution of the species is then forecast using the future climate pattern (subject to the caveats of uncertainty inherent within the GCM scenarios). This provides a map of the expected distribution of the species, given the distribution of suitable climatic conditions for the species at a given future date (e.g., Berry et al. 2002).

However, there may be some serious problems associated with this type of modeling. It has been criticized for its lack of biotic realism, in that it fails to account for the potential impact of several important processes: dispersal ability, evolutionary responses, and biotic interactions (Davis et al. 1998; Pearson and Dawson 2003, 2004; Hampe 2004; Guisan et al. 2006). For example, although suitable climatic conditions might still exist for a given species under future climate scenarios, that species might not be able to disperse at a sufficient rate to track its "climate window." During the process of climate change, the species might also be evolving, possibly developing an altered climatic tolerance that renders the fitted distribution-climate regression inaccurate. Finally, the current range margins of a species might be strongly regulated by biotic interactions; if these interactions break down (or if new interactions are generated, as in the case of the pine processionary moth discussed in Section 6.3.2.2), then the relationship between distribution and climate might also collapse.

If we look specifically at plant–plant interactions, there are many studies that demonstrate the important influence of plant–plant interactions in regulating the distribution of species' ranges. To give just a few examples, Vetaas (2002) demonstrated how the range of temperatures under which Himalayan rhododendron species were capable of growing was far wider when they were grown in ornamental gardens and arboreta, away from the interactions present in their native habitats. Leathwick and Austin (2001) demonstrated how the presence of Nothofagus species in New Zealand old-growth forest strongly modified the climate-distribution fit of other tree species, and Halloy and Mark (2003) showed that removal of low-altitude competitors has enabled some alpine and subalpine species to grow at lower altitude in New Zealand alpine systems. Prince and Carter (1985) examined the survival of individuals of prickly lettuce *Lactuca serriola* outside of its natural range within the UK. They found that there were no sizable direct negative effects of climate on individual plants located beyond the northerly range margin, and concluded that in this instance there was likely to be a strong impact of intraspecific metapopulation-level interactions on the location of the range margin. Furthermore, I have already discussed how facilitation might expand the range margins of some species, particularly into severe environments, where they would be unable to grow in the absence of facilitator species (e.g., Choler, Michalet, and Callaway 2001; Kikvidze et al. 2001; Michalet et al. 2006). These examples all show the potentially strong influence of competitive interactions on distributional limits.

Simple individual-based modeling studies can also demonstrate the potential impact of interactions on the ability of species to track their climate envelopes. Brooker et al. (2007), using a model that was previously developed to look at the spatial arrangement of species along environmental severity gradients (Travis, Brooker, and Dytham 2005; Travis et al. 2006), demonstrate that interactions influence the initial arrangement of species within a landscape, which then, in conjunction with differences in dispersal ability between species, regulates the outcome of altered environmental conditions. Kefi et al. (2007) utilized individual-based models to examine the potential for using changes in vegetation spatial pattern as an early-warning signal of environmental degradation (desertification). They found that the inclusion of facilitative plant interactions in the model was essential for creating realistic spatial pattern dynamics. Broennimann et al. (2007) found that the climate "niche" of the spotted knapweed *Centaurea maculosa* differed substantially between its native European range and its nonnative range in North America, where it is a significant invasive, and proposed that some of this difference may be a consequence of changes in the biotic interactions experienced by the knapweed.

All of these modeling studies, in conjunction with the examples from field-based research, provide a strong argument for the necessity of incorporating plant interactions when modeling climate impacts on species ranges. Unfortunately, however, we are currently in a situation where both sides in the debate still have a case to argue. Although it can be demonstrated that, in some cases, plant–plant interactions might strongly influence the distribution of species, that consideration of such interactions would be essential to effectively model future ranges, and that such strong interactions can be extremely species-specific (Callaway 1998), it might still be said that these occurrences are rare, and that the majority of interactions are generic, i.e., will occur irrespective of whichever species combinations occur in newly formed communities, and will have relatively little impact on the overall distribution of a species, modifying it only at a local, landscape scale (Huston 2002).

However, resistance to the inclusion of interactions in modeling approaches might not only result from the perception that interactions are important only at a local scale, but also from the perceived difficulty of incorporating them into models—a consequence of the apparent species—and context-specificity of interactions. However, as discussed elsewhere in this book, it may be possible to find underlying patterns of where, when, and how interactions play an important role in regulating the establishment, survival, and growth of plant species, and hence the composition of plant communities. Understanding such patterns has been identified as a key gap in population and community ecology (Agrawal et al. 2007) and, with respect to the issue of modeling climate change impacts, might provide a method by which interactions and other biotic processes could be integrated into modeling approaches. In the next section, I will therefore examine a few examples of this type of integrative pattern that already exist within plant community ecology, describing them only briefly but concentrating instead on their relevance to predicting the impacts of climate change on plant community composition.

6.5   WHY PATTERN IS IMPORTANT

6.5.1   Plant Functional Types

From a predictive viewpoint, plant functional types are an attractive concept: If the responses of functional types to particular drivers are consistent, then it is possible to extrapolate an understanding of the impact of environmental drivers from one community to another without the need for detailed knowledge of responses at the species level (Chapin et al. 1996). The relevance of functional types to prediction of climate-change impacts on biodiversity is widely acknowledged (Grime 1997). They have a central role in the development of digital global vegetation models (DGVMs; Woodward and Cramer 1996), and many experimental studies have attempted to determine common responses of functional types to climate change manipulations (e.g., Chapin et al. 1996; Arft et al. 1999; Cornelissen et al. 2001; Dormann and Woodin 2002; Brooker and Van der Wal 2003).

Functional types might also provide us with an approach for integrating the impacts of interactions into climate envelope modeling. Although functional types are commonly associated with life-forms, genuine functionality refers to the processes that a species undertakes (Woodward and Cramer 1996), and the interaction properties associated with a species, e.g., competitor, facilitator, or comparative neutrality, are suitable ways of describing functionality. The broad-scale functional types (strategies) of competitor, stress tolerator, and ruderal (Grime 1974) are linked to underlying differences in competitive ability. On the basis of this categorization, it is clear that competitive functional types dominate in more productive environments, whereas less competitive stress-tolerant species dominate in more severe environments. Changes in climate will therefore be associated with a switch in dominance from stress-tolerants to competitors, and an associated increase in the level of competition in communities.

6.5.2   THE STRESS–GRADIENT HYPOTHESIS

The stress–gradient hypothesis (SGH) has clear direct relevance to understanding the impact of climate change on plant communities. The hypothesis maintains that the dominant type of interaction is influenced by the severity of (or stress imposed by) the environment (e.g., Choler, Michalet, and Callaway 2001; Callaway et al. 2002; Lortie et al. 2004; Callaway 2007; Brooker et al. 2008), and so changes in environmental severity, for example through climate change, will lead to *predictable* changes in the dominant type of interaction occurring within a community. Based on current understanding we would predict, at a very basic level, that in relatively productive environments, competition should be included as a factor potentially limiting plant success (and hence the potential for survival of different growth forms, e.g., competitive and stress tolerant, could be weighted accordingly), while in unproductive environments, relatively stress-sensitive species would be less likely to survive in the absence of stress-tolerant facilitators.

With respect to the SGH, there is currently a need for further exploration of the shape and generality of the relationship between interactions and environmental severity, as considerable debate has taken place concerning the applicability of the SGH in arid or semi-arid environments (see Chapter 7, this volume; Maestre and Cortina 2004; Maestre, Valladares, and Reynolds 2005, 2006; Lortie and Callaway 2006; Brooker et al. 2008). Furthermore, as climate change involves both changes in temperature and precipitation, we must consider under which circumstances the stress–gradient relationship is determined more strongly by temperature or water. As pointed out by Kawai and Tokeshi (2007), it is important to consider the possibility of different relationships between stress and interactions for different environmental drivers. In arctic/alpine systems, temperature alone may be the key issue, and so a direct link between interactions and climate may be easier to determine than in warm, low-latitude environments where the combined effect of high temperatures and low water availability may be crucial. In these lower latitude environments it may be more complicated to predict a general response of interactions to climate change.

6.5.3   THE IMPORTANCE AND INTENSITY OF INTERACTIONS

Closely linked to the SGH, and the ecological theory that underlies it, are the concepts of competition importance and competition intensity. Put simply, the intensity of competition is the absolute reduction in success of a target plant caused by competition, whereas the importance of competition is this decrease expressed as a proportion of the total impact of all environmental factors on target plant success. Making a clear distinction between these two concepts is important in resolving a debate that has been under way in plant ecology for many years, namely whether the role of interactions in plant communities decreases with increasing environmental severity. These concepts were originally outlined by Welden and Slauson (1986), and although there have been calls for their use (Grace 1991), this debate is still promoted because they are not widely recognized (Brooker et al. 2005; Brooker and Kikvidze 2008).

To greatly simplify the debate, there is evidence to suggest that although interactions may be as intense in more severe environments, they are relatively less important. Such an outcome can arise through a similar impact of competition in both stressful and productive environments, but a greater additional abiotic environmental impact in the severe environments. Recent empirical work has demonstrated that this effect may be operating as expected in the field (Gaucherand, Liancourt, and Lavorel 2006). In the context of finding pattern within the role of plant interactions that can be fed into predictive modeling, this would indicate that there is relatively less need to consider the potential impact of interactions in more physically stressful environments. In such systems, there would be a greater direct regulatory impact of the abiotic environmental conditions and hence a better match between the predictions of the climate envelope modeling and the future realized distribution (assuming of course that dispersal limitation and evolution have no impact!).

6.5.4 THE HUMP-BACKED DIVERSITY MODEL

The hump-backed diversity model is one of the classic models of plant community ecology. Originally outlined by Grime (1973), it postulates that plant diversity is highest in those sites with intermediate levels of stress and disturbance. In more severe environments (i.e., more stressed or disturbed), diversity is thought to decline because fewer species are adapted to deal with the environmental conditions. In more productive environments, diversity is also thought to decline but, in this case, through an increased impact of competitive exclusion (although it should be recognized that other mechanisms have been proposed to explain the hump-backed relationship, e.g., Hodgson 1987; Foster 2001).

Recent studies of severe environments have shown the important role that facilitation can play in regulating community composition (Valiente-Banuet et al. 2006; Lortie 2007; Brooker et al. 2008). This has prompted some researchers to attempt to integrate facilitation into the hump-backed model (Hacker and Gaines 1997) and has prompted a recent proposal for a revision of the original model to explicitly include the role of biotic interactions in regulating both sides of the hump-backed relationship (see Chapter 4, this volume; Michalet et al. 2006).

The hump-backed model again indicates that competition is likely to be an important community structuring process in productive environments, with relatively less impact of competition but an increasing role for facilitation in severe ecosystems. Therefore, according to this model, changes in climate that regulate the productivity of the environment will influence the role of interactions in structuring plant communities.

It is clear that these concepts are very closely linked and interdependent. For example, the competitive functional type is likely to be an evolved response to the increased importance of competition within more productive systems, and its presence is in turn related to the regulation of biodiversity in productive ecosystems through competition. Similarly the hump-backed model of species diversity, and the changing role of interactions along productivity gradients, is clearly closely related to the SGH and the debate concerning the importance and intensity of competition. Because of the close links between these concepts, research that helps elucidate one of these concepts will be directly beneficial in helping develop the others. Furthermore, exploration of the links between these topics highlights potential areas for future research. For example, although there may be a switch in the dominant type of interaction from competition in productive systems to facilitation in severe environments, does this necessarily mean that facilitation is as important in structuring communities in severe environments as competition is in productive ecosystems? Could, for example, facilitation drive the evolution of plant functional type in a similar manner to the development of competitive growth forms in productive systems?

Finally, it should be recognized that, in all the cases identified here, the predictions provided concerning how, where, and when plant–plant interactions are important in regulating the composition of plant communities are relatively simplistic and broadscale. However, they represent, at the very least, a starting point for developing a framework that enables better inclusion of interactions into modeling approaches.

6.6   CONCLUSIONS

In this chapter I have discussed, using some general examples, the links between climate change, plant–plant interactions, and the composition of plant communities. I have shown how plant–plant interactions, both facilitative and competitive, are one of the processes by which climate change can influence plant community composition. This can occur through changes in the relative competitive dominance of species within the community, perhaps through changes in phenology or the removal or imposition of abiotic constraints on growth, or the influx or loss of competitor or facilitator species from the community in response to climate. I have also discussed how plant species' ranges can be strongly regulated by plant–plant interactions, and consequently that interactions should be included in models that are attempting to predict the response of species' distributions to climate change. I have then outlined a number of concepts that might be developed into a framework to make such integration more straightforward.

It is, however, essential to point out that climate change is not occurring in isolation. Out of necessity I have discussed it in this very simple manner, but in reality climate change will be occurring in conjunction with other anthropogenic environmental changes, including land-use change, atmospheric nitrogen deposition, the increased frequency of aggressive alien species, and enhanced atmospheric carbon dioxide concentration. All of these factors may influence the composition of plant communities, and in all cases these impacts can be mediated by plant–plant interactions (Brooker 2006). However, even the simplistic examination of the links between climate change and plant–plant interactions presented here has highlighted some clear goals for future research.

It is clearly worth pursuing in more detail the questions that arise when we start to link the various concepts concerning when, where, and how plant–plant interactions play important roles in plant communities. Key questions that we might examine are:

1. What is the shape of the relationship between environmental severity and interactions that underlies the stress–gradient hypothesis? Is this shape generic, or does it vary between systems?
2. Are facilitative interactions as important in severe environments as competitive interactions are in productive environments? Might facilitative interactions have an evolutionary impact and lead to the development of a facilitator functional type?

Undertaking such work would not only help us to predict the response of plant species and community composition to climate change, but would also further our understanding of the fundamental ecological processes that underlie such responses.

ACKNOWLEDGMENTS

I would like to thank Jack Lennon and anonymous referees for very helpful comments on earlier versions of this chapter; Hans Cornelissen for the excellent photographs used in Figure 6.1; and Chris Lortie and Oikos for permission to use Figure 6.2. I am also grateful to the Scottish Government Rural and Environment Research and Analysis Directorate (RERAD) for funding during the production of this chapter.

REFERENCES

Agrawal, A. A., D. D. Ackerly, F. Adler, A. E. Arnold, C. Cáceres, D. F. Doak, E. Post, et al. 2007. Filling key gaps in population and community ecology. *Frontiers in Ecology and the Environment* 5: 145–152.

Arft, A. M., M. D. Walker, J. Gurevitch, J. M. Alatalo, M. S. Bret-Harte, M. Dale, M. Diemer, et al. 1999. Responses of tundra plants to experimental warming: Meta-analysis of the international tundra experiment. *Ecological Monographs* 69: 491–511.

Bellingham, P. J., and D. A. Coomes. 2003. Grazing and community structure as determinants of invasion success by Scotch broom in a New Zealand montane shrubland. *Diversity and Distributions* 9: 19–28.

Berry, P. M., T. P. Dawson, P. A. Harrison, and R. G. Pearson. 2002. Modelling potential impacts of climate change on the bioclimatic envelope of species in Britain and Ireland. *Global Ecology and Biogeography Letters* 11: 453–462.

Bertness, M. D., and R. Callaway. 1994. Positive interactions in communities. *Trends in Ecology and Evolution* 9: 191–193.

Bertness, M. D., and P. J. Ewanchuk. 2002. Latitudinal and climate-driven variation in the strength and nature of biological interactions in New England salt marshes. *Oecologia* 132: 392–401.

Blasing, T. J., and K. Smith. 2008. *Current greenhouse gas concentrations*. Oak Ridge, TN, USA: US Carbon Dioxide Information Analysis Center. http://cdiac.ornl.gov/pns/current_ghg.html.

Broennimann, O., U. A. Treier, H. Müller-Schärer, W. Thuiller, A. T. Peterson, and A. Guisan. 2007. Evidence of climatic niche shift during biological invasion. *Ecology Letters* 10: 701–709.

Brooker, R. W. 2006. Plant–plant interactions and environmental change. *New Phytologist* 171: 271–284.

Brooker, R. W., and T. V. Callaghan. 1998. The balance between positive and negative plant interactions and its relationship to environmental gradients: A model. *Oikos* 81: 196–207.

Brooker, R., and Z. Kikvidze. 2008. Importance: An overlooked concept in plant interaction research. *Journal of Ecology* 96: 703–708.

Brooker, R., Z. Kikvidze, F. I. Pugnaire, R. M. Callaway, P. Choler, C. J. Lortie, and R. Michalet. 2005. The importance of importance. *Oikos* 109: 63–70.

Brooker, R. W., F. T. Maestre, R. M. Callaway, C. L. Lortie, L. A. Cavieres, G. Kunstler, P. Liancourt, et al. 2008. Facilitation in plant communities: The past, the present, and the future. *Journal of Ecology* 96: 18–34.

Brooker, R. W., D. Scott, S. C. F. Palmer, and E. Swaine. 2006. Transient facilitative effects of heather on Scots pine along a grazing disturbance gradient in Scottish moorland. *Journal of Ecology* 94: 637–645.

Brooker, R. W., J. M. J. Travis, E. J. Clark, and C. Dytham. 2007. Modelling species' range shifts in a changing climate: The impacts of biotic interactions, dispersal distance and the rate of climate change. *Journal of Theoretical Biology* 245: 59–65.

Brooker, R., and R. Van der Wal. 2003. Can soil temperature direct the composition of high arctic plant communities? *Journal of Vegetation Science* 14: 535–542.

Brooker, R., and J. Young. 2006. Climate change and biodiversity in Europe: A review of impacts, policy responses, gaps in knowledge and barriers to the exchange of information between scientists and policy makers. Department for Environment, Food and Rural Affairs, London.

Bruno, J. F., J. J. Stachowicz, and M. D. Bertness. 2003. Inclusion of facilitation into ecological theory. *Trends in Ecology and Evolution* 18: 119–125.

Buckley, L. B., and J. Roughgarden. 2004. Effects of changes in climate and land use. *Nature* 430: U2.

Callaghan, T. V., S. Jonasson, and R. W. Brooker. 1997. Arctic clonal plants and global change. In *The ecology and evolution of clonal plants*, ed. H. de Kroon and J. van Groenendael, 381–403. Leiden: Backhuys Publishers.

Callaway, R. M. 1995. Positive interactions among plants. *Botanical Review* 61: 306–349.

Callaway, R. M. 1998. Are positive interactions species-specific? *Oikos* 82: 202–207.

Callaway, R. M. 2007. *Positive interactions and interdependence in plant communities*. Dordrecht, The Netherlands: Springer.

Callaway, R. M., R. Brooker, P. Choler, Z. Kikvidze, C. J. Lortie, R. Michalet, L. Paolini, et al. 2002. Positive interactions among alpine plants increase with stress. *Nature* 417: 844–847.

Callaway, R. M., and F. I. Pugnaire. 1999. Facilitation in plant communities. In *Handbook of functional plant ecology*, ed. F. I. Pugnaire and F. Valladares, 623–648. New York: Marcel Dekker.

CBD. 2003. Secretariat of the Convention on Biological Diversity. Interlinkages between biological diversity and climate change. Advice on the integration of biodiversity considerations into the implementation of the United Nations Framework Convention on Climate Change and its Kyoto Protocol. CBD Technical Series No. 10. SCBD, Montreal.

Chapin, F. S. I., M. S. Bret-Harte, S. E. Hobbie, and H. Zhong. 1996. Plant functional types as predictors of transient responses of arctic vegetation to global change. *Journal of Vegetation Science* 7: 347–358.

Choler, P., R. Michalet, and R. M. Callaway. 2001. Facilitation and competition on gradients in alpine plant communities. *Ecology* 82: 3295–3308.

Cornelissen, H. C., T. V. Callaghan, J. M. Alatalo, A. Michelsen, E. Graglia, A. E. Hartley, D. S. Hik, et al. 2001. Global change and arctic ecosystems: Is lichen decline a function of increase in vascular plant biomass? *Journal of Ecology* 89: 984–994.

Davis, A. J., L. S. Jenkinson, J. H. Lawton, B. Shorrocks, and S. Wood. 1998. Making mistakes when predicting shifts in species range in response to global warming. *Nature* 391: 783–786.

Dormann, C. F., and S. J. Woodin. 2002. Climate change in the Arctic: Using plant functional types in a meta-analysis of field experiments. *Functional Ecology* 16: 4–17.

Dormann, C. F., R. Van der Wal, and S. J. Woodin. 2004. Neighbour identity modifies effects of elevated temperature on plant performance in the High Arctic. *Global Change Biology* 10: 1587–1598.

Dunnett, N. P., and J. P. Grime. 1999. Competition as an amplifier of short-term vegetation responses to clime: An experimental test. *Functional Ecology* 13: 388–395.

EEA. 2004. European Environment Agency. Impacts of Europe's changing climate. An indicator-based assessment. Copenhagen.

Emmett, B. A., C. Beier, M. Estiarte, A. Tietema, H. L. Kristensen, D. Williams, J. Peñuelas, I. Schmidt, and A. Sowerby. 2004. The response of soil processes to climate change: Results from manipulation studies of shrubland across an environmental gradient. *Ecosystems* 7: 625–637.

Fitter, A. H., and R. S. R. Fitter. 2002. Rapid changes in flowering time in British plants. *Science* 296: 1689–1691.

Foster, B. L. 2001. Constraints on colonization and species richness along a grassland productivity gradient: The role of propagule availability. *Ecology Letters* 4: 530–535.

Foster, B. L., T. L. Dickson, C. A. Murphy, I. S. Karel, and V. H. Smith. 2004. Propagule pools mediate community assembly and diversity-ecosystem regulation along a grassland productivity gradient. *Journal of Ecology* 92: 435–449.

Gaucherand, S., P. Liancourt, and S. Lavorel. 2006. Importance and intensity of competition along a fertility gradient and across species. *Journal of Vegetation Science* 17: 455–464.

Grabherr, G., M. Gottfried, and H. Pauli. 1994. Climate effects on mountain plants. *Nature* 369: 448.

Grace, J. B. 1991. A clarification of the debate between Grime and Tilman. *Functional Ecology* 5: 583–587.

Grime, J. P. 1973. Competitive exclusion in herbaceous vegetation. *Nature* 242: 344–347.

Grime, J. P. 1974. Vegetation classification by reference to strategy. *Nature* 250: 26–30.

Grime, J. P. 1979. *Plant strategies and vegetation processes.* Chichester: Wiley.

Grime, J. P. 1997. Climate change and vegetation. In *Plant ecology*, ed. M. J. Crawley, 582–594. Oxford: Blackwell Publishing.

Grime, J. P. 1998. Benefits of plant diversity to ecosystems: Immediate, filter and founder effects. *Journal of Ecology* 86: 902–910.

Guisan, A., A. Lehmann, S. Ferrier, M. Austin, M. C. Jacob, C. Overton, R. Aspinall, and T. Hastie. 2006. Making better biogeographical predictions of species' distributions. *Journal of Applied Ecology* 43: 386–392.

Hacker, S. D., and S. D. Gaines. 1997. Some implications of direct positive interactions for community species diversity. *Ecology* 78: 1990–2003.

Halloy, S. R., and A. F. Mark. 2003. Climate-change effects on alpine plant biodiversity: A New Zealand perspective on quantifying the threat. *Arctic, Antarctic and Alpine Research* 35: 248–254.

Hampe, A. 2004. Bioclimate envelope models: What they detect and what they hide. *Global Ecology and Biogeography Letters* 13: 469–470.

Harte, J., A. Ostling, J. L. Green, and A. Kinzig. 2004. Climate change and extinction risk. *Nature* 430: U3.

Harte, J., and R. Shaw. 1995. Shifting dominance within a montane vegetation community: Results of a climate-warming experiment. *Science* 267: 876–880.

Heegaard, E., and V. Vandvik. 2004. Climate change affects the outcome of competitive interactions: An application of principal response curves. *Oecologia* 139: 459–466.

Hódar, J. A., and R. Zamora. 2004. Herbivory and climatic warming: A Mediterranean outbreaking caterpillar attacks a relict, boreal pine species. *Biodiversity and Conservation* 13: 493–500.

Hodgson, J. G. 1987. Why do so few species exploit productive habitats? An investigation into cytology, plant strategies and abundance within a local flora. *Functional Ecology* 1: 243–250.

Honnay, O., K. Verheyen, J. Butaye, H. Jacquemyn, B. Bossuyt, and M. Hermy. 2002. Possible effects of habitat fragmentation and climate change on the range of forest plant species. *Ecology Letters* 5: 525–530.

Hubbell, S. P. 2001. *The unified neutral theory of biodiversity and biogeography*. Princeton, NJ: Princeton University Press.

Huston, M. A. 2002. Introductory essay: Critical issues for improving predictions. In *Predicting species occurrences: Issues of accuracy and scale*, ed. J. M. Scott, P. J. Heglund, M. L. Morrison, J. B. Haufler, M. G. Raphael, W. A. Wall, and F. B. Samson, 7–21. Covelo, CA: Island Press.

IPCC. 2007. Climate change 2007: Synthesis report. An assessment of the Intergovernmental Panel on Climate Change. IPCC, United Nations, New York.

Kawai, T., and M. Tokeshi. 2007. Testing the facilitation-competition paradigm under the stress-gradient hypothesis: Decoupling multiple stress factors. *Proceedings of the Royal Society of London Series B (Biological Sciences)* 274: 2503–2508.

Kéfi, S., M. Rietkerk, C. L. Alados, Y. Pueyo, V. P. Papanastasis, A. ElAich, and P. C. de Ruiter. 2007. Spatial vegetation pattern and imminent desertification in Mediterranean arid ecosystems. *Nature* 449: 213–217.

Kikvidze, Z., L. Khetsuriani, D. Kikodze, and R. M. Callaway. 2001. Facilitation and interference in subalpine meadows of the central Caucasus. *Journal of Vegetation Science* 12: 833–838.

Klanderud, K. 2005. Climate change effects on species interactions in an alpine plant community. *Journal of Ecology* 93: 127–137.

Klein, J. A., J. Harte, and X.-Q. Zhao. 2004. Experimental warming causes large and rapid species loss, dampened by simulated grazing, on the Tibetan Plateau. *Ecology Letters* 7: 1170–1179.

Kullman, L. 2001. 20th century climate warming and tree-limit rise in the Southern Scandes of Sweden. *Ambio* 30: 72–80.

Laakso, J., V. Kaitala, and E. Ranta. 2001. How does environmental variation translate into biological processes? *Oikos* 92: 119–122.

Leathwick, J. R., and M. P. Austin. 2001. Competitive interactions between tree species in New Zealand's old-growth indigenous forests. *Ecology* 82: 2560–2573.

Lortie, C. J. 2007. An ecological tardis: The implications of facilitation through evolutionary time. *Trends in Ecology and Evolution* 22: 627–629.

Lortie, C. J., R. W. Brooker, P. Choler, Z. Kikvidze, R. Michalet, F. I. Pugnaire, and R. M. Callaway. 2004. Rethinking plant community theory. *Oikos* 107: 433–438.

Lortie, C. J., and R. M. Callaway. 2006. Re-analysis of metaanalysis: Support for the stress-gradient hypothesis. *Journal of Ecology* 94: 7–16.

Maestre, F. T., and J. Cortina. 2004. Do positive interactions increase with abiotic stress? A test from a semi-arid steppe. *Proceedings of the Royal Society of London Series B, Supplement* 271: S331–S333.

Maestre, F. T., F. Valladares, and J. F. Reynolds. 2005. Is the change of plant–plant interaction with abiotic stress predictable? A meta-analysis of field results in arid environments. *Journal of Ecology* 93: 748–757.

Maestre, F. T., F. Valladares, and J. F. Reynolds. 2006. Does one model fit all? A reply to Lortie and Callaway. *Journal of Ecology* 94: 17–22.

McGill, B. J. 2003. A test of the unified neutral theory of biodiversity. *Nature* 422: 881–885.

Michalet, R., R. W. Brooker, L. A. Cavieres, Z. Kikvidze, C. J. Lortie, F. I. Pugnaire, A. Valiente-Banuet, and R. M. Callaway. 2006. Do biotic interactions shape both sides of the humped-back model of species richness in plant communities? *Ecology Letters* 9: 767–773.

Millennium Ecosystem Assessment. 2005. *Ecosystems and human well-being: Biodiversity synthesis*. Washington DC: World Resources Institute.

Molau, U., and J. M. Alatalo. 1998. Responses of subarctic-alpine plant communities to simulated environmental change: Biodiversity of bryophytes, lichens and vascular plants. *Ambio* 27: 322–329.

Morecroft, M. D., G. J. Masters, V. K. Brown, I. P. Clarke, M. E. Taylor, and A. T. Whitehouse. 2004. Changing precipitation patterns alter plant community dynamics and succession in an ex-arable grassland. *Functional Ecology* 18: 648–655.

Nee, S., and G. Stone. 2003. The end of the beginning for neutral theory. *Trends in Ecology and Evolution* 18: 433–434.

Parmesan, C. 2006. Ecological and evolutionary responses to recent climate change. *Annual Review of Ecology, Evolution and Systematics* 37: 637–669.

Parmesan, C., and G. Yohe. 2003. A globally coherent fingerprint of climate change impacts across natural systems. *Nature* 421: 37–42.

Pearson, R. G., and T. P. Dawson. 2003. Predicting the impacts of climate change on the distribution of species: Are bioclimate envelope models useful? *Global Ecology and Biogeography*: 12: 361–371.

Pearson, R. G., and T. P. Dawson. 2004. Bioclimate envelope models: What they detect and what they hide. Response to Hampe. *Global Ecology and Biogeography Letters* 13: 471–473.

Peñuelas, J., and I. Filella. 2001. Responses to a warming world. *Science* 294: 793–795.

Pigott, C. D., and J. P. Huntley. 1981. Factors controlling the distribution of *Tilia cordata* at the northern limits of its geographical range, III: Nature and causes of seed sterility. *New Phytologist* 87: 817–839.

Press, M. C., T. V. Callaghan, and J. A. Lee. 1998. How will European arctic ecosystems respond to projected global environmental change? *Ambio* 27: 306–311.

Preston, C. D., M. G. Telfer, H. R. Arnold, and P. Rothery. 2002. The changing flora of Britain. In *New atlas of the British and Irish flora*, ed. C. D. Preston, D. A. Pearman, and T. D. Dines, 35–35. Oxford: Oxford University Press.

Prince, S. D., and R. N. Carter. 1985. The geographical distribution of prickly lettuce (*Lactuca serriola*), III: Its performance in transplant sites beyond its distribution limit in Britain. *Journal of Ecology* 73: 49–64.

Rao, S. J., G. R. Iason, I. A. R. Hulbert, D. A. Elston, and P. A. Racey. 2003. The effect of sapling density, heather height and season on browsing by mountain hares on birch. *Journal of Applied Ecology* 40: 626–638.

Root, T. L., J. T. Price, K. R. Hall, S. H. Schneider, C. Rosenzweig, and A. Pounds. 2003. Fingerprints of global warming on wild animals and plants. *Nature* 421: 57–60.

Rousset, O., and J. Lepart. 1999. Shrub facilitation of *Quercus humilis* regeneration in succession on calcareous grasslands. *Journal of Vegetation Science* 10: 493–502.

Rousset, O., and J. Lepart. 2000. Positive and negative interactions at different life stages of a colonizing species (*Quercus humilis*). *Journal of Ecology* 88: 401–412.

Schwartz, M. D., R. Ahas, and A. Aasa. 2006. Onset of spring starting earlier across the northern hemisphere. *Global Change Biology* 12: 343–351.

Sparks, T. H., and A. Menzel. 2002. Observed changes in seasons: An overview. *International Journal of Climatology* 22: 1715–1725.

Sturm, M., C. Racine, and K. Tape. 2001. Increasing shrub abundance in the Arctic. *Nature* 411: 546.

Takenaka, A. 2005. Local coexistence of tree species and the dynamics of global distribution pattern along an environmental gradient: A simulation study. *Ecological Research* 20: 297–304.

Tans, P. 2008. NOAA/ESPL(www.esrl.noaa.gov/and/ccgg/trends/).

Tape, K., M. Sturm, and C. Racine. 2006. The evidence for shrub expansion in northern Alaska and the Pan-Arctic. *Global Change Biology* 12: 686–702.

Thomas, C. D., A. Cameron, R. E. Green, M. Bakkenes, L. J. Beaumont, Y. C. Collingham, B. F. N. Erasmus, et al. 2004. Extinction risk from climate change. *Nature* 427: 145–148.

Thomas, C. D., and J. L. Lennon. 1999. Birds extend their ranges northwards. *Nature* 399: 213.

Thuiller, W., M. B. Araújo, R. G. Pearson, R. J. Whittaker, L. Brotons, and S. Lavorel. 2004. Uncertainty in predictions of extinction risk. *Nature* 430: U1–U1.

Travis, J. M. J. 2002. Climate change and habitat destruction: A deadly anthropogenic cocktail. *Proceedings of the Royal Society of London B: Biological Sciences* 270: 467–473.

Travis, J. M. J., R. W. Brooker, E. J. Clark, and C. Dytham. 2006. The distribution of positive and negative species interactions across environmental gradients on a dual-lattice model. *Journal of Theoretical Biology* 241: 896–902.

Travis, J. M. J., R. W. Brooker, and C. Dytham. 2005. The interplay of positive and negative species interactions across an environmental gradient: Insights from an individual-based simulation model. *Biology Letters* 1: 5–8.

Valiente-Banuet, A., A. V. Rumebe, M. Verdú, and R. M. Callaway. 2006. Quaternary Plant lineages sustain global diversity by facilitating Tertiary lineages. *Proceedings of the National Academy of Sciences USA* 103: 16812–16817.

Vetaas, O. R. 2002. Realized and potential climate niches: a comparison of four rhododendron tree species. *Journal of Biogeography* 29: 545–554.

Walker, M. D., C. H. Wahren, R. D. Hollister, G. H. R. Henry, L. E. Ahlquist, J. M. Alatalo, M. S. Bret-Harte, et al. 2006. Plant community responses to experimental warming across the tundra biome. *Proceedings of the National Academy of Sciences USA* 103: 1342–1346.

Walther, G.-R. 2003. Plants in a warmer world. *Perspectives in Plant Ecology, Evolution and Systematics* 6: 169–185.

Walther, G.-R., S. Berger, and M. T. Sykes. 2005. An ecological "footprint" of climate change. *Proceedings of the Royal Society of London B: Biological Sciences* 272: 1427–1432.

Walther, G.-R., E. Post, P. Convey, A. Menzel, C. Parmesan, T. J. C. Beebee, J.-M. Fromentin, O. Hoegh-Guldberg, and F. Bairlein. 2002. Ecological responses to recent climate change. *Nature* 416: 389–395.

Watson, R. T. 2005. Turning science into policy: Challenges and experiences from the science-policy interface. *Philosophical Transactions of the Royal Society of London B: Biological Sciences* 360: 471–477.

Welden, C. W., and W. L. Slauson. 1986. The intensity of competition versus its importance: An overlooked distinction and some implications. *The Quarterly Review of Biology* 61: 23–44.

Woodward, F. I., and W. Cramer. 1996. Plant functional types and climatic change: Introduction. *Journal of Vegetation Science* 7: 306–308.

Zhou, L., C. J. Tucker, R. K. Kaufmann, D. Slayback, N. V. Shabanov, and R. B. Myneni. 2001. Variations in northern vegetation activity inferred from satellite data of vegetation index during 1981 to 1999. *Journal of Geophysical Research* 106: 20069–20083.

Zobel, M. 1997. The relative role of species pools in determining plant species richness: An alternative explanation of species coexistence? *Trends in Ecology and Evolution* 12: 266–269.

7 Synthetic Analysis of the Stress–Gradient Hypothesis

Christopher J. Lortie

CONTENTS

7.1 INTRODUCTION

In this chapter, the merit or quality of a single, specific hypothesis is critically evaluated using the stress–gradient hypothesis. While this is clearly not the most precise hypothesis, it is commonly used as an heuristic to interpret changes in plant–plant interactions at various points on gradients/continuums and, sometimes, on contiguous ranges within populations or communities. In it simplest form, the stress–gradient hypothesis predicts that the relative frequency of positive interactions between plants will increase with increasing consumer pressure or environmental stress (Bertness and Callaway 1994). This initial formulation is a hypothesis that generated (directly and indirectly) an extensive set of testable predictions, including the following:

The sign of net interactions will change with the environment, typically gradients (Brooker and Callaghan 1998; Callaway et al. 2002; Holmgren, Scheffer, and Huston 1997; Holzapfel and Mahall 1999; Pugnaire and Luque 2001).

Positive interactions are species specific (Aksenova and Onipchenko 1998; Callaway 1998).

Life stage can shift the outcome of net interactions (Lortie and Turkington 2002; Tielborger and Kadmon 2000).

Changing spatial scale similarly shifts the net interactions (Lortie et al. 2005; Malkinson, Kadmon, and Cohen 2003; Sthultz, Gehring, and Whitham 2007).

The measured response can change the estimation of net interactions (Kawai and Tokeshi 2007; Travis et al. 2006).

Importance and intensity do not necessarily covary positively (Brooker and Kikvidze 2008; Brooker et al. 2005).

A comprehensive review of positive interactions details future directions within this subdiscipline (Brooker et al. 2008). However, there is some confusion due to the interpretations of importance versus intensity versus frequency of occurrence on gradients (Brooker et al. 2005; Gaucherand, Liancourt, and Lavorel 2006), which has only recently been tested directly (Lamb and Cahill 2008), nor is the hypothesis always necessarily supported empirically (Maestre, Valladares, and Reynolds 2005). For the purposes of the treatment discussed herein and as a logical starting point, the stress–gradient hypothesis is best considered in its principal form. To clarify anew, the *frequency* of positive interactions increases at relatively more environmentally limiting locations within a defined spatial extent, such as along a continuous gradient within a single habitat. Frequency is the relative occurrence of a net interaction with a given sign, not intensity or importance; facilitation is a positive sign of the net interaction between two or more species detected experimentally or mensuratively via association; and in this context, gradient refers to different positions within a given habitat, often designated as low versus high stress. The scope of inference of this hypothesis should thus be appropriately limited to the frequency of sign changes at different positions in an environment. Hence, a clear discussion of the key elements of gradients and stress is provided in this chapter, using this context with more-advanced dialogues on the topic best reserved for the primary literature, as indices, experiments, and our understanding of the relative importance of net interactions continue to evolve. Nonetheless, it is always useful to revisit assumptions of the hypothese extended in the literature.

The content in this chapter will not focus on review per se but, rather, more on the strength of the associated logic, concepts, and a subset of the published empirical tests. Admittedly, narrative reviews, vote counting, and comprehensive lists of independent experimental approaches are compelling, but there are other means available for ecologists to explore the relative strength of a hypothesis (Gurevitch and Hedges 2001), including formal meta-analyses comparing effects sizes between studies (Gates 2002), systematic reviews (Higgins and Green 2006; Pullin and Stewart 2006), and models (Anderson and Burnham 2001; Wiegert 1988). To an extent, these approaches have been previously applied to the stress–gradient hypothesis, including meta-analyses (Lortie and Callaway 2005; Maestre, Valladares, and Reynolds 2005), systematic reviews (Flores and Jurado 2003), and conceptual modeling of some aspects of stress and facilitation (Lortie et al. 2004a, 2004b; Michalet et al. 2006), which in all cases have refined our understanding of gradients and interactions. The analyses presented in this chapter are novel and are used to make general recommendations on how to potentially progress within this subdiscipline of experimental plant ecology. The approach applied to develop these ideas is as follows: a history of gradients and how gradients are used, followed by an exploration of stress, and then finally a combination of both these ideas with an analysis of empirical tests for the stress–gradient hypothesis. Best practices for application and reporting of gradient studies are also proposed. Conceptual models and graphics are used where possible to articulate general attributes of gradients and stress for the benefit of the reader.

7.2 PURPOSE

This chapter is presented in an effort to explore the importance of abiotic gradients in understanding plant–plant interactions, patterns in diversity, and individual plant-level responses to variation in the environment. More specifically, the purpose is to explore what has become a reasonably dominant paradigm in plant ecology: that differences in "stress" are an important organizing concept for predicting the outcome of plant–plant interactions at different points in space and time.

7.3 A BRIEF HISTORY OF GRADIENTS IN ECOLOGY: ALL THE WORLD IS A GRADIENT

7.3.1 Definition of an Ecological Gradient

We use four levels of derivation when we assume a gradient is present in plant ecology. Using first principles, however, a gradient in its simplest sense is the change in a suite of abiotic parameters on a surface as it moves through space (and perhaps time). More proximately, an ecological gradient is typically assumed within a study when a change in the organism(s) of interest varies as one changes position, i.e., less productivity moving in one direction or a change in species composition, or when the environmental factors seem to change (preferably measured), i.e., less moisture, lower temperature, etc. A gradient in plant ecology is therefore composed of directional, spatially consistent changes in (a) the vegetation structure or composition and (b) underlying environmental parameters. The necessary condition for an ecological gradient is, however, the directional change in the underlying variation in parameter(s), which is likely quite complex, and it is assumed that it can be simplified to a set of several important, preferably causal, factors. Typically, gradients are not inferred for either monocultures (natural or otherwise) or for mosaics but, rather, for gradual changes likely varying in a continuous manner in space. However, direct correspondence between vegetation and the environment is not a necessary condition for a set of samples to be designated as a gradient. In what is arguably the first formal discussion of gradients, direct and indirect gradients are delineated wherein vegetation arranges or sorts according to position or not, respectively (Whittaker 1967). Using the terminology of vector calculus, we can identify the four levels of assumptions we make as plant ecologists as follows.

7.3.2 DEGREE OF EXTRACTION

1. Point processes are sampled or assumed on a physical surface in an intuitive manner (i.e., increased elevation up a mountain or away from the water's edge in intertidal environments).
2. These points are summed to a scalar surface of variation in the abiotic parameter set, and it is generally recognized as a dimensional multivariate response surface.
3. Both biotic and abiotic vectors are imposed on this on a scalar surface of variation.
4. Each set of vectors is then summed and simplified to a trend analysis or a single, possibly representative vector.

This set of simplifying assumptions is a necessary means to organize the complexity present in any natural system, and it allows plant ecologists to identify points to sample, to infer process, and to organize patterns described or observed at various levels of organization. Typically, the approach intuitively used by ecologists parallels the derivation of a gradient in vector calculus. More simply and directly, environmental variables are observed or measured; these measures are summarized or integrated to form a sense of directionality in the abiotic parameters; both changes in the vegetation and the environment are compared; and in tests for the relative frequency of net interactions, a trend analysis is derived to estimate the outcome of interactions between individuals relative to other points in space within this particular set of samples (Figure 7.1). Thus, at each position in space along the inferred gradient, x_0, x_1, to x_n, there is a potential vector of change such as decreasing temperature, moisture, etc., in the abiotic variables. Often, it is assumed that each of these environmental vectors moves in concert, and this can be tested using multivariate statistics (Manly 1994; McCune and Medford 1999). This set of steps is the most common approach applied to virtually all published studies testing the stress–gradient hypothesis. Unfortunately, however, the multivariate analyses are not commonly applied to the underlying environmental data sets to ascertain whether the gradients are direct or indirect (Pausas and Austin 2001).

It is useful to deconstruct the assumptions made when inferring a gradient for at least two fundamental reasons. Firstly, by explicitly identifying how gradients are defined and described in plant ecology studies testing for net interactions, we can increase the level of circumspection or critical appraisal associated with each step to ensure that others can evaluate the relative strength of each assumption. Additionally, this transparency would promote better data reporting practices and encourage the application of a more quantitative approach (such as multivariate analyses) to the assignment of position on a gradient rather than an arbitrary low- versus high-stress designation. Oddly, there is a disconnect between the rigorous pattern description common in vegetation science and the tests for net interactions in plant ecology (Austin 1999). In the latter instance, the simplifying assumption of stress on gradients is as a

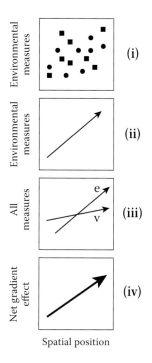

FIGURE 7.1 The four steps commonly used in describing (or sometimes assuming) a gradient for the purposes of testing for changes in the outcome of net interactions between plants. Respectively, environmental attributes are observed or measured, environmental directionality is identified, vegetation and environmental constraints are imposed, and the pattern in both environment and vegetation is simplified and then measured to a degree decided by the researcher. Points shown in panel (i) denote two different possible measures observed, such as moisture or temperature. Arrows in panels (ii) and (iii) denote vectors that assume directional, consistent change in the point processes summarized. Width of vectors denotes degree of inference, i.e., wider = greater degree of extraction, and the labels "e" and "v" refer to environmental and vegetation patterns, respectively. Spatial position is the potential positions sampled within a given contiguous habitat. Commonly, a gradient is defined as a discrete, continuous habitat, such as a connected alpine meadow on a mountain or a single uninterrupted sand dune community moving from an ocean ending at a forest.

simple linear vector applied to the design and implementation of most experimental tests of this hypothesis. Secondly, errors associated with assumptions at one level can be propagated to other levels of inference, thereby reducing both (a) the predictive capacity of a model or hypothesis such as the stress–gradient hypothesis and (b) the ability to identify data gaps or lack of support, i.e., a predicted shift is not detected because of failure to meet underlying, often implicit, assumptions, and this does not reflect the validity of the model. To better test the stress–gradient hypothesis, we must therefore inspect each assumption associated with defining both gradients and stress to ensure that we are effectively testing this hypothesis.

It goes without saying that we often make these ecological gradient derivations with very little or in some cases no data, and instead we rely heavily on our intuitive

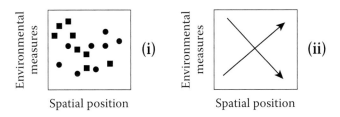

FIGURE 7.2 Environmental measures do not necessarily move in concert within a given spatial context. In this instance, a true gradient is present. However, causal factors do not have the same effect on the net plant interactions. Each symbol denotes a different abiotic factor.

sense of natural history and observation of a natural system. However, it is also reasonable to assume that an error in assumptions or incomplete/inaccurate data at one level could be propagated to higher levels of numerical averaging, thereby reducing our ability to understand and predict plant community responses. Several simple classes of specific errors to the stress–gradient hypothesis are thus possible and include the following:

The scalar field of environmental data is variable, not consistent (not all vectors have the same slope). We often only sample two points and designate as "higher" or "lower" relative stress with little data (potential lack of a true gradient).

More than two points are sampled, but the full range of environmental variation is not sampled (veiled gradients). Therefore, the type of gradient and response variables measured may be limited and fail to capture the community dynamics.

In the first instance, the assumption that the scalar field is consistent is not met, i.e., potentially causal environmental variables do not covary in concert (Figure 7.2). The implication of this violation is that the net effect of the gradient can have variable or contrasting effects on community dynamics (Michalet 2007). Commonly, positions on gradients are arbitrarily assigned to relatively high or low stress (Figure 7.3)

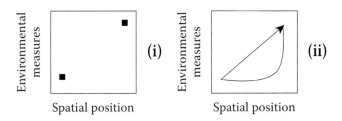

FIGURE 7.3 Two points in space do not necessarily constitute a gradient or allow assessment of the rate of change in the abiotic parameter set at different positions within the habitat or the defined sample space. The linear vector shows the commonly assumed change in the environmental measure along the inferred gradient, but the parameter may change nonlinearly with position (curved line).

without clear quantification of abiotic parameters (Lortie and Callaway 2005), and this can either be incorrect or misleading, since the definition of two points does not allow (a) assessment of the variation in net interactions with changes in the continuous underlying parameter set (Kawai and Tokeshi 2007) or (b) assessment of whether the changes in the parameters monotonically vary with positions.

Observation or measurement of abiotic parameters at several points on an inferred gradient is desirable and enhances the potential validity of the assumptions used in delineating a gradient. Unfortunately, however, it is also possible that an inadequate range is sampled or, in other words, that the estimate of the gradient length is cropped. This error can lead to "veiled gradients" (McCoy 2002), in that any local gradient sampled over a short period of time or across a small geographic range will reveal only a portion of its true variation, leaving the remainder of the "true" or actual gradient veiled in space or time, thereby reducing our capacity to infer its importance in shifting net interactions between plants (Figure 7.4). This error can be particularly troublesome in that several points are measured and directionality identified, but if the vegetation or net interactions do not change linearly with change in position on the gradient, then the arbitrary range selected may not capture the shift in net interactions. Hence, the researcher is erroneously led to conclude that the gradient is unimportant or that the hypothesis is unsupported.

Finally, not all gradients or species are the same. This may seem trivial and obvious, but it is not an uncommon error to overlook these differences. In reviewing an excellent empirical study that addresses this limitation with an appropriately structured experimental design incorporating different species, experimental treatments, and scales (Sthultz, Gehring, and Whitham 2007), Michalet (2007) accurately describes the error common in this literature (but not in the study reviewed) as failure to identify and differentiate between functional strategies of interacting species and the type of constraint changing within a given gradient. The functional characteristics and limiting factors should thus be carefully and systematically selected or sampled as diversely as possible. Gradients are real phenomena in natural systems and provide an invaluable opportunity and tool to test for net interactions.

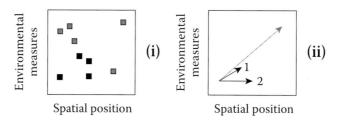

FIGURE 7.4   An example of veiled gradients wherein the true range of variation is not sampled within a particular gradient, leaving a proportion of the potentially causal variation undetermined. Filled symbols (in black) show sampled variation, and unfilled symbols (in gray) show actual range of environmental variation. In this example, the dark vector denotes the inferred vector from the data sampled (labeled 1), the gray vector denotes the actual gradient from the entire extent of data available, and they both share similar attributes. In a more extreme erroneous instance, however, one could envisage the dark vector with a zero slope (labeled 2) and the gray vector a positive slope, leading to an incorrect conclusion that there is no gradient.

While this discussion of errors may seem extensive, the potential benefits of sampling communities or structuring experiments to capitalize on observed variation to whatever degree possible (preferably being more quantitative) generally far outweigh the pitfalls associated with deriving a gradient (see Section 7.4). Furthermore, gradients provide a means to organize significant variation in space and time and thereby infer a degree of sorting within communities attributable to ecophysiological fit and plant–plant interactions.

7.4   HOW GRADIENTS ARE USED, TESTED, AND QUANTIFIED

Proximately, we move through space or time to sample variation or replicate an experiment to estimate the effects of the environment on a given plant community. The aim of this section is to (a) provide the reader with a brief list of representative tangible examples of gradients as a hypothesis-testing tool, (b) introduce the concept of gradient length, which is important to the stress–gradient hypothesis, and (c) emphasize that species may respond differently to gradients and that this difference provides an opportunity to test for net interactions.

7.4.1   GRADIENTS AS A HYPOTHESIS-TESTING TOOL

Plant ecology is an extensive and highly productive discipline, and literally thousands of studies have shown that both positive and negative interactions are frequent and sometimes important (Brooker et al. 2005; Brooker and Kikvidze 2008; Callaway 2007; Goldberg et al. 1999). It is worthwhile to briefly consider the full range of gradients as a tool to test hypotheses regarding interactions, including the stress–gradient hypothesis (Table 7.1). Ultimately, in all instances gradients were used to infer relevance of a hypothesis through assessment of changes in interactions using the environmental variation inherent in the natural system. As developed in the previous section, there are caveats associated with appropriately defining a gradient. However, it is evident through inspection of the examples listed that gradients effectively serve as a means to explore shifts in the net interactions between plants and that the product of these shifts can be used to explain numerous attributes of a plant community. In short, the hypotheses tested via gradients differ in that the community attribute predicted and described changes, but central to virtually every example is the consideration of net interactions. As such, the detailed consideration of the stress–gradient hypothesis in this chapter is an excellent representative case study of gradient usage that may not necessarily be unique in its limitations and strengths.

7.4.2   THE CONCEPT OF GRADIENT LENGTH

As described in Section 7.3, gradients are a means to organize information and patterns in environmental variation within an environment. As such, a gradient is a tool that compares relative differences in measures and responses within a specific, a priori, defined context by the researcher. This delineation within a particular context, i.e., habitat, should be as broad as possible, but the choice to sample either particular locations or inherent differences in the total variation within a particular habitat

TABLE 7.1

List of Sample Publications That Use Gradients as a Tool to Test a Wide Variety of Concepts in Plant Ecology

Concept and Example

Gradients and competition (Gaucherand, Liancourt, and Lavorel 2006)

Gradients and density dependence (Stevens and Carson 1999)

Gradients and dispersal (Svenning and Skov 2002)

Gradients and diversity (Hacker and Gaines 1997)

Gradients and facilitation (Callaway et al. 2002)

Gradients and herbivory (Olofsson 2001)

Gradients and invasion (Lortie and Cushman 2007)

Gradients and restoration (Gomez-Aparicio et al. 2004)

Gradients and seed bank dynamics (Cavieres and Arroyo 2001)

Gradients and species pool (Schamp, Laird, and Aarssen 2002)

Note: The concepts listed are very general and summarize only the class or set of hypotheses that may be present within each particular subdomain, i.e., there are many hypotheses tested with respect to patterns in diversity using gradients, but with only one example at the conceptual level listed herein. The citations listed were selected based on their clarity in using a gradient to explore a particular concept; in all instances, the listed citation is a good starting point for additional examples within that particular set of studies.

generates variation in the length of gradients sampled. Gradient length is a key consideration with respect to the stress–gradient hypothesis and is best defined as the spatial extent of the environmental variation sampled within a particular system (Kawai and Tokeshi 2007; Lortie and Callaway 2005; McCoy 2002). Hence, even studies within similar habitats may differ in estimates of net interactions between individuals due to differences in the range of abiotic variation sampled (Figure 7.5). As such, it is important when comparing the relative importance of a particular study—either formally via meta-analyses or informally via description or narrative reviews—to assess the length of each gradient sampled.

One potential means to do this would be to compare the absolute difference in the effect sizes of the vegetation using measures such as LRR (log response ratio), RNE (relative neighbor effect), or RII (relative interaction index) (Armas, Ordiales, and Pugnaire 2004; Goldberg et al. 1999; Hedges, Gurevitch, and Curtis 1999; Markham and Chanway 1996) at the two most extreme points sampled within each study (Figure 7.5). Even more directly, assessment of the limitations associated with each position on the gradient would be another means to extend inference beyond directionality of gradient to magnitude of difference. Another alternative solution is to control for differences between studies by including latitude, mean annual precipitation, and estimates of aridity or limitation, and then using these designations as a means to assign potential positions on gradients or assess gradient length. Admittedly, sites at similar latitudes or with similar estimates of mean annual precipitation can vary dramatically in the degree to which gradients are manifested therein and, as

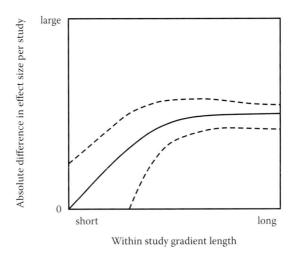

FIGURE 7.5   The importance of variation in gradient length within a study and the absolute difference in mean effect size per study. The within-study gradient length could be calculated by taking the difference in the stress index for the two most extreme points along the gradient tested in a study. The absolute difference in mean effect size is the difference in the effect sizes of a response variable at various points tested within a study. The dotted lines depict a representative 95% C.I.

such, composite measures of environmental impact or gradient length are preferable to single-point estimates of environmental gradient attributes (Parker et al. 1999). In summary, not all gradients are created equal, and it is prudent to quantify as thoroughly as possible the degree of difference within the gradient in a given study and control for differences between gradients when comparing studies.

7.4.3   The Importance of Species-Specific Responses to Gradients

Not only can differences between gradients be an important consideration, but within gradients, differences between species can influence both assessment of the impact of the gradient and the outcome of net interactions. This issue has been explored in several contexts, including the importance of species identity as nurse plants (Callaway 1998), functional type of the species tested (Sthultz, Gehring, and Whitham 2007), and importance of abiotic factor dependence on the species (Kawai and Tokeshi 2007). The sign and strength of species interactions can vary linearly and nonlinearly on gradients (Kawai and Tokeshi 2007), and thus differences between species can introduce significant differences in the capacity of a gradient to test hypotheses. For instance, by ignoring interactions between species and more simply modeling the environmental sensitivity of species to gradients, it is probable that species may have the same values in some estimate of performance, such as biomass or survival, at two different points on a gradient but differ dramatically in how they "get to" each of those points (Figure 7.6). This is an intriguing conceptual notion, in that the

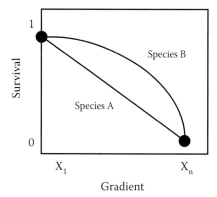

FIGURE 7.6 Species-specific responses to a gradient may be nonlinear and dissimilar. In this instance, species A and B both have similar measures of survival at two different points on a gradient, but may have very different sets of responses to other points on the gradient.

degree of divergence within a gradient between individual species responses may also vary nonlinearly, thereby generating a range of differences for selection, niche partitioning, and net interactions to act upon (Figure 7.7).

Adapting the terminology of selection theory (Gould and Lewontin 1979) (see Figure 7.8):

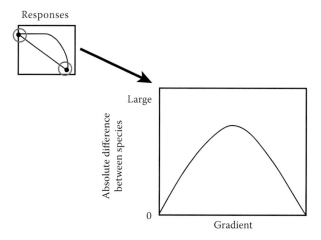

FIGURE 7.7 Differences in the shape of the response by species to a gradient may differ, be nonlinear, and generate patterns of divergence that can act as a substrate for higher-order processes. The panel in the top left of this figure is the response of two species to a gradient (independent axis is gradient, dependent axis is a performance measure, sensu Figure 7.6). This difference can be revisualized as the absolute difference in a performance measure at every point on a gradient as the lower panel in this figure. At the lower left of the dependent axis, no difference in performance is thus zero, and large differences are depicted as the hump in this plot.

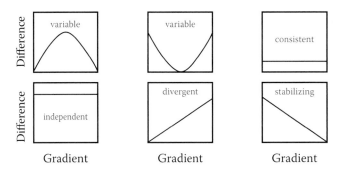

FIGURE 7.8 The absolute differences between two species on a gradient. Differences in this instance refer to the absolute difference in any performance measure such as biomass, survival, reproduction, etc., between two species measured at the same points on a spatial gradient along the entire extent of the gradient. The terms in gray on each plot refer to categorizations of the potential differences in the gradient responsiveness of species from independent to variable.

> Differences between the responsiveness of species to gradients can be variable (or perhaps disruptive), whereby the directionality of a gradient ultimately influences each species at the extremes, but the responses vary at every other position.
>
> Differences between species responsiveness to the gradient may always be large with no slope, suggesting species independence.
>
> Differences may be minimal, approaching zero, thereby suggesting that species respond consistently to the gradient.
>
> Species differences may either increase or decrease monotonically (divergent and stabilizing responsiveness, respectively).

In the latter two instances in particular, we might predict that increasing differences between species along a gradient provides opportunity for species that are more different to interact either more positively (Grime 1977) or more negatively (Tilman 1988), depending on the paradigm that one subscribes to (Vazquez and Stevens 2004).

This conceptual modeling exercise of species responsiveness to gradients highlights several proximate consideration when using gradients to test hypotheses, including the stress–gradient hypothesis. Differences are most likely nonlinear, and this is an opportunity ecologically to better understand community construction and interactions. However, sampling more extensively along a gradient with the same measures is more important than sampling more intensively but to a lesser spatial extent.

7.5 STRESS AS SHORTHAND FOR ENVIRONMENTAL VARIATION: SHOULD WE BE STRESSED ABOUT USING STRESS?

The concept of *stress* is often criticized as ecological quackery or unscientific, since it is both ubiquitous and broadly applied to different phenomena. Nonetheless, the clearest statement of stress that has been articulated in the literature is that it is the sum of external constraints limiting the rates of resource acquisition, growth, or reproduction of organisms (Grime 1989). Hence, "[a] world of infinitely large populations is impossible because most environments can support only limited numbers" (White 2001). Different environments can support different numbers of individuals, and one way these differences might be manifested is through variation in the severity of particular environments in time or space. However, severity or stress (like limitation) is not a simple, singular concept (Körner 2003; Lortie et al. 2004b). Stress encompasses multiple levels of organization and scales and has been applied to some of the most fascinating phenomena in ecology, including different physical environments.

As Grime proposed, stress is best considered as an external force with attributes such as its nature, severity, or periodicity (1989). This theme is reiterated in many other seminal publications that consider the utility of the concept of stress by exploring its various attributes (Calow 1989; Parker et al. 1999; Underwood 1989). A major concern has emerged, however, since it is both a simple definable concept as proposed above and a conceptual construct that subsumes and embodies other ecological concepts and attributes, in that it has been applied very informally to describe a wide range of phenomena. This is not necessarily a problem. Unfortunately in ecology, we do not often use techniques associated with information science in spite of the fact that many, if not most, of the concepts we deal with are pluralistic and explicitly quantitative.

A common tool used in information science to handle such concepts is an ontology that formally represents a set of related concepts within a specific context (Bard and Rhee 2004; Gruber 1993; Michener and Brunt 2000). Here, a short synthesis of attributes included in the concept of stress is delineated using an ontology to further clarify the specific usage of this term.

Stress is often used in a quick and dirty fashion, often in an "always or never" fashion (Körner 2003). However, this provides us with a challenge to refine and improve usage rather than dismiss. Both stress and its dependent defining concept limitation are accurately defined (and have been for some time), but their descriptive applications are not. *Limitation* is a useful concept when applied to individuals within a species but not at greater levels, i.e., the community (Lortie et al. 2004a, 2004b). However, measurement of changes in the limitation of a species in the *context of interactions with other species* is a highly interesting avenue of experimentation, particularly in light of the large-scale anthropogenic changes in multiple factors that may have previously served as limitations for many species, i.e., increased length of growing season and temperature. Stress is a useful concept when considering changes both within and between species at different points in space and time. While it is convenient to refer to environments as stressful, more specific descriptions of the abiotic parameters being referred to should accompany this general label and will likely reduce the bias against its use as an adjective to describe extreme environments or stress at this scale.

Perhaps the best example of an effective application of the concept of stress is a recent paper evaluating the relative importance of disturbance and stress in predicting species diversity for 14 different alpine plant communities in the Alps (Kammer and Mohl 2002). Thirteen different factors potentially limiting biomass production were tested in addition to six types of disturbance. While the data were primarily survey-based using a ranking of degree of limitation of a given factor's ability to reduce biomass, 10 of the 14 alpine plant communities tested were generally controlled by stress factors and not by disturbance (Kammer and Mohl 2002). This is an excellent application of the concept of stress. Admittedly, it is tempting to argue that whether stress was a good predictor or not is irrelevant, since it is based upon the opinion of scientists working in that region and not on exact quantification. However, it does strongly suggest that ecologists do use stress as an organizational concept. Furthermore, there is nothing stopping us from using this general concept to screen which parameters to measure in the field (perhaps all 13), or from using a composite index for stress or even a global model to integrate all the abiotic measures available (Kammer and Mohl 2002). If one of the goals of ecology is to predict the distribution and abundance of not only individual species in space and/or time but changes in diversity of different community types, then using general concepts such as stress as an initial measure to explore patterns can be a powerful explanatory technique. This is not to say that it is always necessary to do so, but that in some cases we can work from broad concepts to more reductionistic measures—perhaps meeting those who approach problems from the more reductionistic levels somewhere in the middle— thereby generating common causal models.

At a conceptual level, the attributes of stress such as nature, severity, and period- icity (Grime 1989)—or at the population level, inertia, resilience, stability—along with timing, magnitude, and order of stresses should also be used to specifically describe what is being tested even for a single abiotic factor (Underwood 1989). While there may be "no single best hierarchical level at which to study the effect of stress" (Parker et al. 1999), we must begin to specify the scope of usage of the terms and concepts associated with stress and clearly discuss it within explicit contexts. Furthermore, delineating whether a proximate or ultimate effect of stress is detected within a study would necessarily clarify the scope of the conclusions that might be derived from a given study, i.e., does the reduction in biomass change the fitness of individual species or the diversity of species within the community (Kammer and Mohl 2002). In our experiments, we do "need to be alert for the presence of these deviations" (Körner 2003) from optimality and the many solutions that exist to cope with them, including genetics *and* interactions with other species and the specific attributes of stress that might be concomitantly changing.

To this end, an ontology is proposed herein for the general concept of stress that encompasses an appropriate set of related concepts at all levels of organization that are applied to the description and interpretation of stress. (See Figure 7.9, derived from the collection of papers published in 1989 in the *Biological Journal of the Linnaean Society*, vol. 37.) This depiction of the concept of stress, which includes the different attributes of stress, the different types of effects on individuals, and the responses that might occur at different levels of organization, demonstrates that, while it is a very general concept, it can have very specific and testable applications in many aspects of ecological research. The characteristics, scale, and choice of species strongly influence the predictability of detecting stress (Parker et al. 1999), and specification of scope within studies will enhance our capacity to generalize relevance of this concept.

Most elements of this ontology are very clear. However, it should be emphasized that there is a significant difference between the potential proximate and ultimate effects of stress on an individual. The proximate effect is the immediate effect of an environmental stress on an individual, and the ultimate is the evolution of a strategy to

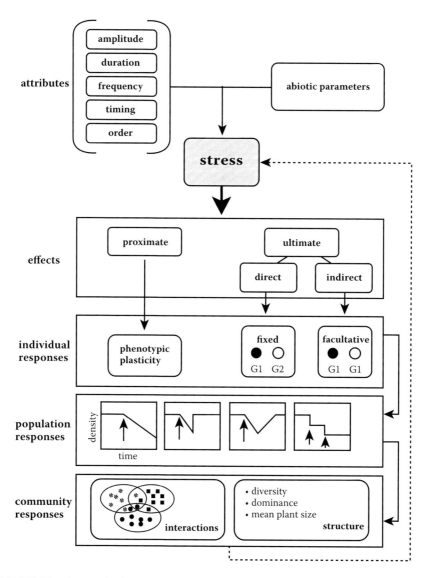

FIGURE 7.9 An ontological synthesis of the ideas and concepts used throughout the literature to describe stress at all levels of organization (concepts derived from a collection of papers published in 1989 in the *Biological Journal of the Linnaean Society*, vol. 37). Stress is a complex concept with different attributes that influence the abiotic factors (or the suite thereof) that can be viewed as stress, depending on their effect on the organism. The negative effects can include proximate and ultimate (both direct and indirect), which are classified according to the response of the individual. The responses include phenotypic plasticity, fixed changes depending on genotype (depicted as G1 or G2), or facultative (same genotype, G1, but different phenotypic responses—depicted by open and closed circles). Population-level responses to stress can also vary, depending primarily on how the density of the population recovers following the stress event(s), and community-level responses can vary, depending on interactions between individuals of different species or in terms of community structure.

deal with this stress (Calow 1989). Or put another way, there can be a negative effect on the organism (such as reduced height or biomass) but no direct effect on fitness (seed production or recruitment). It is also possible for genotypes of the same species to respond (or cope) in different ways by evolving different strategies to deal with the same abiotic stress (Calow 1989), or for there to be direct and indirect effects of stress on the organism, depending on the life stage (Sibly and Calow 1989). Additionally, it does not necessarily follow that all species *must or can* respond to stress in a fixed fashion leading to the evolution of either stress tolerance or avoidance. Some species may be unable to evolve the traits necessary to reduce the effects of a particular stress within a habitat, either due to genetic constraints (i.e., lack of appropriate genetic variability, including unfavorable linkages or pleiotropy), or due to trade-offs between defense, survival, or growth (Bradshaw and Hardwick 1989; Sibly and Calow 1989), or simply due to limitation in the environment to be able to do so (Bryant, Chapin, and Klein 1983). In contrast, other species or even genotypes within a species may respond facultatively through phenotypic plasticity (Calow 1989). Virtually all aspects of the concept stress within this visual definition have been studied in plant ecology, but it is clear from inspection of the literature that it is easy to confuse level of organization and characteristics, since the relationship between the different levels at which stress can be considered has not been linked (until now).

7.6   A META-ANALYTICAL TEST OF THE TOP 10 PUBLISHED STUDIES OF THE STRESS–GRADIENT HYPOTHESIS

Systematic reviews and formal meta-analyses are increasingly useful synthetic tools for ecologists to assess the generality and validity of hypotheses (Gates 2002). Commonly, these approaches use a broad, clearly defined search criterion to restrict the set of studies used in the synthetic endeavor. In other words, the scope of inference is defined by the studies selected and is transparent to the reader. To date, two publications have directly assessed the generality of the importance of the stress–gradient hypothesis, but in both instances the scale was very broad across all arid and semi-arid environments, and in some respects, generality of conclusions was substituted for precision, particularly in the latter publication (Flores and Jurado 2003; Maestre and Cortina 2004). Another example successfully applied a formal meta-analysis to the stress–gradient hypothesis but within a more restrictive context (Gomez-Aparicio et al. 2004). In this application, the studies were replicated experiments in the alpine with the same experimental approach, and the meta-analysis detected a consistent and significant shift from negative to positive within increasing stress. Here, a common ground is proposed in the spirit of synthesis and as a starting point for the reader to consider the general utility of the stress–gradient hypothesis.

The top 10 studies testing the stress–gradient hypothesis were selected using the following criteria: compelling evidence for support of the hypothesis, clear and unequivocal differences in stress on the gradient used, and the survival data necessary for a meta-analysis reported in the published study (Table 7.2). Approximately 25 studies met these criteria (Lortie and Callaway 2005), although less rigid criteria would generate a longer, but more controversial list of potential studies for synthesis (Maestre and Cortina 2004). Sensitivity analysis via repeated selection with resampling of studies

TABLE 7.2
Arbitrary Set of the Top 10 Studies Testing the Stress–Gradient Hypothesis in Plant Ecology

Experimental Approach and Study	Functional Type of Nurse Plant
Manipulative (Callaway et al. 2002)	forb
Manipulative (Chambers 2001)	tree
Observational (García-Fayos and Gasque 2002)	shrub
Observational (Greenlee and Callaway 1996)	forb
Manipulative (Ibanez and Schupp 2001)	shrub
Manipulative (Maestre 2002)	shrub
Manipulative (Maestre et al. 2001)	shrub
Manipulative (Peek and Forseth 2003)	shrub
Observational (Sthultz, Gehring, and Whitham 2007)	tree
Manipulative (Valiente-Banuet and Ezcurra 1991)	succulent

Note: Entries were selected from a list of 25 studies that clearly tested the stress–gradient hypothesis. For full details of the statistics and approach associated with a similar version of this data set, see Lortie and Callaway (2005). Two studies are added from previous work, including Callaway et al. (2002) and Sthultz, Gehring, and Whitham (2007). In the latter instance, both experimental and observational data were available, but only the observational data reported variances and so was selected here. In the former study, a mean of two sites was used where survival was accurately recorded.

was done, and results reported are robust. Hence, this is not a biased study selection for two reasons. First, resampling did not affect the outcome, and second, the primary purpose of this review was to ascertain whether the stress–gradient can effectively predict patterns when effectively applied. By way of analogy using evidence-based medicine, the question here is whether a certain drug can effectively treat a disease versus a much broader review that would explore the specific set of conditions when it might work best or whether it is effective in all instances. All appropriate meta-analytic criteria were fulfilled (Gates 2002), including the loading of only one data point per level per study; 95% confidence intervals are reported (generated via bootstrapping with 9999 iterations); a fixed-effects model was fit, since the scope of inference is this particular set of studies; and publication bias and heterogeneity were assessed using the approaches used in previous studies with MetaWin 2.1.5 (Lortie and Callaway 2005; Maestre and Cortina 2004; Rosenberg, Adams, and Gurevitch 2000).

In this synthetic analysis, the top 10 studies included a wide range of plant functional types and included studies from alpine, arid, and semi-arid systems. Survival data was tested using the effect size metric LnOR (natural log of relative odds ratio). Given the lack of axillary data, weighted regression analyses to compare relative stress between sites or assess differences in gradient lengths were not possible.

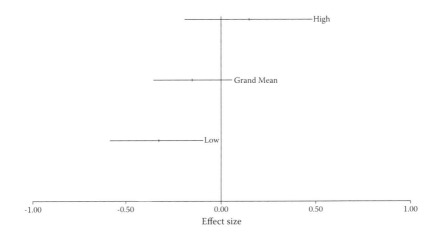

FIGURE 7.10 Mean survival effect estimates and sign for a meta-analysis of the top 10 studies testing the stress–gradient hypothesis. Error bars denote 95% confidence intervals. Effect size is lnOR (natural log of odds ratio), and "low" versus "high" refers the relative stress positions within each gradient tested. Negative effect sizes represent reduced survival, and positive denotes increased survival with neighbors or dominant plant (i.e., nurse plant) present, depending on the particular study.

Hence, three key statistical comparisons were done. Comparison of sign at high versus low stress, comparison of mean differences, and comparison of differences in heterogeneity were explored. Using the top 10 stress–gradient hypothesis studies, a shift from negative to positive survival in the presence of neighbors/nurse plants at low versus high stress was detected (Figure 7.10). This clearly supports the primary predictive element of the stress–gradient hypothesis—namely that the frequency of positive interactions increases with increasing stress within a gradient.

However, as previously discussed, assessment of mean differences is very difficult and likely inappropriate using a meta-analysis with this type of data due to the increased variation associated with mixing gradients of different lengths, which is uncontrolled for here (Lortie and Callaway 2005). Not surprisingly then, neither the grand mean nor the mean effect of neighbors at high elevations was significantly different from a net effect of zero at the 95% confidence level (i.e., C.I. overlap 0 in these two instances, Figure 7.10), and due to the low confidence in this statistical test in comparing mean effects from varying length gradients, it is not clear whether the mean differences are not significantly different or are simply from different points within a distribution. In essence, it is appropriate to compare whether the frequency of positive versus negative net effects is different via meta-analysis of uncontrolled gradient data, but not whether the estimated variation associated with a mean effect overlaps (Lipsey and Wilson 2001). Finally, the magnitude statistical variation between groups was not significantly different from random (Q_b = 3.31, p = .07, df = 1), but the difference within groups was significantly different (Q_w = 226.18, p < .0001, df = 18). This strongly supports the interpretation that patterns in variation associated with sampling arbitrary points reduces the capacity to assess mean effects.

In summary, a meta-analysis of the top 10 stress–gradient studies suggests that this hypothesis effectively predicts a shift from net negative to net positive effects with stress in this set of studies but that gradient length must be reported in future studies to facilitate control for differences in the extent of environmental variation sampled between studies.

7.7 WHY USE GRADIENTS AND HOW TO DO SO: RECOMMENDATIONS FOR STRUCTURING GRADIENT STUDIES

Gradients provide plant ecologists with the opportunity to explore whether differences in plant–plant interactions change with environmental variation. Nonetheless, the conceptual and empirical themes developed in this chapter highlight the importance of using gradients more precisely to test hypotheses. The stress–gradient hypothesis is a useful paradigm applied to the testing and interpretation of plant–plant interactions and has significantly furthered the conceptual and empirical progress of plant ecology (Bruno, Stachowicz, and Bertness 2003). In general, the evidence considered in this chapter supported the stress–gradient hypothesis logically, conceptually, and empirically. Table 7.3 provides the reader with a summary of the best practices

TABLE 7.3

Checklist for Best Practices in Testing and Reporting Stress–Gradient Hypothesis Experiments

Section	Recommendation	Yes/No	Citation
7.3	Ensure that you have a gradient, i.e., test assumptions in deriving gradient		Pausas and Austin (2001)
7.4	Define scope of inference and spatial extent of gradient		Kawai and Tokeshi (2007)
7.4	Report standardized effect size measures with variance of the vegetation at all points sampled on the gradient		Gomez-Aparicio et al. (2004)
7.4	Test hypothesis both manipulatively and mensuratively within the gradient		Sthultz et al. (2007)
7.5	Use multivariate analyses where possible to explore and simply quantify environmental attributes delineated as stress		Parker et al. (1999)
7.5	Record multiple measures of performance		Brooker et al. (2005)
7.5	Consider using phytometers		Michalet (2007)
7.6	Calculate gradient length		Lortie and Callaway (2006)
7.6	Test for sign differences in addition to statistical tests for mean effects		Lortie and Callaway (2005)

proposed herein for the application and reporting of gradient studies and stress in plant ecology.

REFERENCES

Aksenova, A. A., and V. G. Onipchenko. 1998. Plant interactions in alpine tundra: 13 years of experimental removal of dominant species. *Ecoscience* 5: 258–270.

Anderson, D. R., and K. P. Burnham. 2001. Commentary on models in ecology. *ESA Bulletin* 82: 160–161.

Armas, C., R. Ordiales, and F. Pugnaire. 2004. Measuring plant interactions: A new comparative index. *Ecology* 85: 2682–2686.

Austin, M. P. 1999. The potential contribution of vegetation ecology to biodiversity research. *Ecography* 22: 465–484.

Bard, J. B. L., and S. Y. Rhee. 2004. Ontologies in biology: Design, applications, and future challenges. *Nature Reviews* 5: 213–222.

Bertness, M. D., and R. Callaway. 1994. Positive interactions in communities. *Trends in Ecology and Evolution* 9: 191–193.

Bradshaw, A. D., and K. Hardwick. 1989. Evolution and stress: Genotypic and phenotypic components. *Biological Journal of the Linnaean Society* 37: 137–155.

Brooker, R. W., and T. V. Callaghan. 1998. The balance between positive and negative plant interactions and its relationship to environmental gradients: A model. *Oikos* 81: 196–207.

Brooker, R. W., and D. Kikvidze. 2008. Importance: An overlooked concept in plant interaction research. *Journal of Ecology* 96: 703–708.

Brooker, R. W., Z. Kikvidze, F. Pugnaire, R. M. Callaway, P. Choler, C. J. Lortie, and R. Michalet. 2005. The importance of importance. *Oikos* 109: 63–70.

Brooker, R. W., F. T. Maestre, R. M. Callaway, C. J. Lortie, L. A. Cavieres, G. Kunstler, P. Liancourt, et al. 2008. Facilitation in plant communities: The past, present, and the future. *Journal of Ecology* 96: 18–34.

Bruno, J. F., J. J. Stachowicz, and M. D. Bertness. 2003. Inclusion of facilitation into ecological theory. *Trends in Ecology and Evolution* 18: 119–125.

Bryant, J. P., F. S. Chapin, and D. R. Klein. 1983. Carbon/nutrient balance of boreal plants in relation to vertebrate herbivory. *Oikos* 40: 357–368.

Callaway, R. M. 1998. Are positive interactions species-specific? *Oikos* 82: 202–207.

Callaway, R. M. 2007. *Positive interactions and interdependence in plant communities.* Dordrecht, the Netherlands: Springer.

Callaway, R. M., R. W. Brooker, P. Choler, Z. Kikvidze, C. J. Lortie, R. Michalet, L. Paolini, et al. 2002. Positive interactions among alpine plants increase with stress. *Nature* 417: 844–848.

Calow, P. 1989. Proximate and ultimate responses to stress in biological systems. *Biological Journal of the Linnaean Society* 37: 173–181.

Cavieres, L., and M. T. K. Arroyo. 2001. Persistent soil seed banks in *Phacelia secunda* (Hydrophyllaceae): Experimental detection of variation along an altitudinal gradient in the Andes of central Chile (33° S). *Journal of Ecology* 89: 31–39.

Chambers, J. C. 2001. *Pinus monophylla* establishment in an expanding pinon-juniper woodland: Environmental conditions, facilitation and interacting factors. *Journal of Vegetation Science* 12: 27–40.

Flores, J., and E. Jurado. 2003. Are nurse-protege interactions more common among plants from arid environments? *Journal of Vegetation Science* 14: 911–916.

García-Fayos, P., and M. Gasque. 2002. Consequences of a severe drought on spatial patterns of woody plants in a two-phase mosaic steppe of *Stipa tenacissima* L. *Journal of Arid Environments* 52: 199–208.

Gates, S. 2002. Review of methodology of quantitative reviews using meta-analysis in ecology. *Journal of Animal Ecology* 71: 547–557.

Gaucherand, S., P. Liancourt, and S. Lavorel. 2006. Importance and intensity of competition along a fertility gradient and across species. *Journal of Vegetation Science* 17: 455–464.

Goldberg, D. E., T. Rajaniemi, J. Gurevitch, and A. Stewart-Oaten. 1999. Empirical approaches to quantifying interaction intensity: Competition and facilitation along productivity gradients. *Ecology* 80: 1118–1131.

Gomez-Aparicio, L., R. Zamora, J. M. Gomez, J. A. Hodar, J. Castro, and E. Baraza. 2004. Applying plant facilitation to forest restoration: A meta-analysis of the use of shrubs as nurse plants. *Ecological Applications* 14: 1128–1138.

Gould, S. J., and R. C. Lewontin. 1979. The spandrels of San Marco and the Panglossian paradigm: A critique of the adaptationist programme. *Proc. R. Soc. Lond. B* 205: 581–598.

Greenlee, J. T., and R. M. Callaway. 1996. Abiotic stress and the relative importance of interference and facilitation in montane bunchgrass communities in western Montana. *American Naturalist* 148: 386–396.

Grime, J. P. 1977. Evidence for the existence of three primary strategies in plants and its relevance to ecological and evolutionary theory. *American Naturalist* 111: 1169–1194.

Grime, J. P. 1989. The stress debate: Symptom of impending synthesis? *Biological Journal of the Linnaean Society* 37: 3–17.

Gruber, T. R. 1993. Towards principles for the design of ontologies used for knowledge sharing. Technical report KSL 93-04. Knowledge Systems Laboratory, Stanford University, CA.

Gurevitch, J., and L. V. Hedges. 2001. Meta-analysis: Combining the results of independent experiments. In *Design and analysis of ecological experiments*, ed. S. M. Scheiner and J. Gurevitch, 347–369. Oxford: Oxford University Press.

Hacker, S. D., and S. D. Gaines. 1997. Some implications of direct positive interactions for community species diversity. *Ecology* 78: 1990–2003.

Hedges, L. V., J. Gurevitch, and P. Curtis. 1999. The meta-analysis of response ratios in experimental ecology. *Ecology* 80: 1150–1156.

Higgins, J. P. T., and S. Green, eds. 2006. *Cochrane handbook for systematic reviews of interventions*. Ver. 4.2.6. The Cochrane Collaboration. www.cochrane-handbook.org/.

Holmgren, M., M. Scheffer, and M. A. Huston. 1997. The interplay of facilitation and competition in plant communities. *Ecology* 78: 1966–1975.

Holzapfel, C., and B. E. Mahall. 1999. Bidirectional facilitation and interference between shrubs and annuals in the Mojave Desert. *Ecology* 80: 1747–1761.

Ibanez, I., and E. W. Schupp. 2001. Positive and negative interactions between environmental conditions affecting *Cercocarpus ledifolius* seedling survival. *Oecologia* 129: 624–628.

Kammer, P. M., and A. Mohl. 2002. Factors controlling species richness in alpine plant communities: An assessment of the importance of stress and disturbance. *Arctic, Antarctic, and Alpine Research* 34: 398–407.

Kawai, T., and M. Tokeshi. 2007. Testing the facilitation-competition paradigm under the stress–gradient hypothesis: Decoupling multiple stress factors. *Proc. R. Soc. Lond. B* 274: 2503–2508.

Körner, C. 2003. Limitation and stress: Always or never? *Journal of Vegetation Science* 14: 141–143.

Lamb, E. G., and J. F. Cahill. 2008. When competition does not matter: Grassland diversity and community composition. *American Naturalist* 171: 777–787.

Lipsey, M. W., and D. B. Wilson. 2001. *Practical meta-analysis*. Thousand Oaks, CA: Sage Publications.

Lortie, C. J., R. W. Brooker, P. Choler, Z. Kikvidze, R. Michalet, F. Pugnaire, and R. M. Callaway. 2004a. Rethinking plant community theory. *Oikos* 107: 63–70.

Lortie, C. J., R. W. Brooker, Z. Kikvidze, and R. M. Callaway. 2004b. The value of stress and limitation in an imperfect world: A reply to Korner. *Journal of Vegetation Science* 15: 577–580.

Lortie, C. J., and R. M. Callaway. 2005. Re-analysis of meta-analysis: Support for the stress–gradient hypothesis. *Journal of Ecology* 94: 7–16.

Lortie, C. J., and J. H. Cushman. 2007. Effects of a directional abiotic gradient on plant community dynamics and invasion in a coastal dune system. *Journal of Ecology* 95: 468–491.

Lortie, C. J., E. Ellis, A. Novoplansky, and R. Turkington. 2005. Implications of spatial pattern and local density on community-level interactions. *Oikos* 109: 495–502.

Lortie, C. J., and R. Turkington. 2002. The facilitative effects by seeds and seedlings on emergence from the seed bank of a desert annual plant community. *Ecoscience* 9: 106–111.

Maestre, F. T. 2002. La restauración de la cubierta vegetal en función del patrón espacial de factores bióticos y abióticos. Universidad de Alicante, Spain.

Maestre, F. T., S. Bautista, J. Cortina, and J. Bellot. 2001. Potential for using facilitation by grasses to establish shrubs on a semiarid degraded steppe. *Ecological Applications* 11: 1641–1655.

Maestre, F. T., and J. Cortina. 2004. Do positive interactions increase with abiotic stress? A test from a semi-arid steppe. *Proc. Royal Soc. London Suppl.* 271: S331–S333.

Maestre, F. T., F. Valladares, and J. F. Reynolds. 2005. Is the change of plant–plant interactions with abiotic stress predictable? A meta-analysis of field results in arid environments. *Journal of Ecology* 93: 748–757.

Malkinson, D., R. Kadmon, and D. Cohen. 2003. Pattern analysis in successional communities: An approach for studying shifts in ecological interactions. *Journal of Vegetation Science* 14: 213–222.

Manly, B. F. J. 1994. *Multivariate statistical methods: A primer.* 2nd ed. Boca Raton, FL: Chapman and Hall/CRC.

Markham, J. H., and C. P. Chanway. 1996. Measuring plant neighbour effects. *Functional Ecology* 10: 548–549.

McCoy, E. D. 2002. The "veiled gradients" problem in ecology. *Oikos* 99: 189–192.

McCune, B., and M. J. Medford. 1999. Multivariate analysis of ecological data. MJM Software, Gleneden Beach, OR.

Michalet, R. 2007. Highlighting the multiple drivers of change in interactions along stress gradients. *New Phytologist* 173: 3–6.

Michalet, R., R. W. Brooker, L. Cavieres, Z. Kikvidze, C. J. Lortie, F. I. Pugnaire, A. Valiente-Banuet, and R. M. Callaway. 2006. Do biotic interactions shape both sides of the humped-back model of species richness in plant communities? *Ecology Letters* 9: 767–773.

Michener, W. K., and J. W. Brunt. 2000. *Ecological data: Design, management and processing.* Oxford: Blackwell Science.

Olofsson, J. 2001. Influence of herbivory and abiotic factors on the distribution of tall forbs along a productivity gradient: Transplantation experiment. *Oikos* 94: 351–357.

Parker, E. D., V. E. Forbes, C. Ritter, C. Barata, W. Admiraal, L. Levin, V. Loeschke, P. Lyytikainen-Saarenmaa, H. Hogh-Jensen, P. Calow, and B. J. Ripley. 1999. Stress in ecological systems. *Oikos* 86: 179–184.

Pausas, J. G., and M. P. Austin. 2001. Patterns of plant species richness in relation to different environments: An appraisal. *Journal of Vegetation Science* 12: 153–166.

Peek, M. S., and I. N. Forseth. 2003. Microhabitat dependent responses to resource pulses in the aridland perennial, *Cryptantha flava. Journal of Ecology* 91: 457–466.

Pugnaire, F. I., and M. T. Luque. 2001. Changes in plant interactions along a gradient of environmental stress. *Oikos* 93: 42–49.

Pullin, A. S., and G. B. Stewart. 2006. Guidelines for systematic review in conservation and environmental management. *Conservation Biology* 20: 1647–1656.

Rosenberg, M. S., D. C. Adams, and J. Gurevitch. 2000. *Meta Win: Statistical software for meta-analysis.* Ver. 2.1.4. Sinauer Associates, Inc., Sunderland, MA.

Schamp, B. S., R. A. Laird, and L. W. Aarssen. 2002. Fewer species because of uncommon habitat? Testing the species pool hypothesis for low plant species richness in highly productive habitats. *Oikos* 97: 145–152.

Sibly, R. M., and P. Calow. 1989. A life-cycle theory of responses to stress. *Biological Journal of the Linnaean Society* 37: 101–116.

Stevens, M. H. H., and W. P. Carson. 1999. Plant density determines species richness along an experimental fertility gradient. *Ecology* 80: 455–465.

Sthultz, C. M., C. A. Gehring, and T. G. Whitham. 2007. Shifts from competition to facilitation between a foundation tree and a pioneer shrub across spatial and temporal scales in a semiarid woodland. *New Phytologist* 173: 135–145.

Svenning, J. C., and F. Skov. 2002. Mesoscale distribution of understorey plants in temperate forest (Kalo, Denmark): The importance of environment and dispersal. *Plant Ecology* 160: 169–185.

Tielborger, K., and R. Kadmon. 2000. Temporal environmental variation tips the balance between facilitation and interference in desert plants. *Ecology* 81: 1544–1553.

Tilman, D. 1988. *Plant strategies and the dynamics and structure of plant communities.* Princeton, NJ: Princeton University Press.

Travis, J. M. J., R. W. Brooker, E. J. Clark, and C. Dytham. 2006. The distribution of positive and negative species interactions across environmental gradients on a dual-lattice model. *Journal of Theoretical Biology* 214: 896–902.

Underwood, A. J. 1989. The analysis of stress in natural populations. *Biological Journal of the Linnaean Society* 37: 51–78.

Valiente-Banuet, A., and E. Ezcurra. 1991. Shade as a cause of the association between the cactus *Neobuxbaumia tetetzo* and the nurse plant *Mimosa luisana* in the Tehuacan Valley, Mexico. *Journal of Ecology* 79: 961–971.

Vazquez, D. P., and R. D. Stevens. 2004. The latitudinal gradient in niche breadth: Concepts and evidence. *American Naturalist* 164: E1–E19.

White, T. C. R. 2001. Opposing paradigms: Regulation or limitation of populations? *Oikos* 93: 148–152.

Whittaker, R. H. 1967. Gradient analysis of vegetation. *Biological Reviews* 49: 207–264.

Wiegert, R. G. 1988. Holism and reductionism in ecology: Hypotheses, scale and systems models. *Oikos* 53: 267–269.

Index